SAGE was founded in 1965 by Sara Miller McCune to support the dissemination of usable knowledge by publishing innovative and high-quality research and teaching content. Today, we publish over 900 journals, including those of more than 400 learned societies, more than 800 new books per year, and a growing range of library products including archives, data, case studies, reports, and video. SAGE remains majority-owned by our founder, and after Sara's lifetime will become owned by a charitable trust that secures our continued independence.

Los Angeles | London | New Delhi | Singapore | Washington DC | Melbourne

Advance Praise

An abundance of publications have been written on the blueprints of organizational excellence; many of these examples are no longer valid today since quite a few of the mentioned enterprises no longer exist. This book gives a practical insight of a manager who has persevered to make things happen, devising winning strategies into genuine sustained results of two troubled organizations. A must read!

Alberto Hassan
Former CEO Orinoco Iron, Venezuela; Founding President of HBIA & IIMA

One of the rare books on corporate turnaround stories in the Indian ecosystem—must read for practicing managers on how to lead through adversity and break corporate inertia to maintain excellence. First-hand account of the miraculous growth journey and cultural change of mammoth National Mineral Development Corporation (NMDC) and Hindustan Copper Limited (HCL)—both Public Sector Units, make it all the more exciting.

R. K. Sharma
Secretary General, Federation of Indian Mining Industries (FIMI)

A fascinating insight into the challenges in managing diverse companies to success from radically different positions. The author speaks with authority and insight about how to achieve a turnaround in fortunes, not just for a failing company but also for what appeared on the surface to be a successful company.

Mr Jim Lennon
Former Chairman of Commodities Research,
Macquarie Capital (Europe) Ltd

The Sleeping Tigers is a fascinating story about the evolution and turnaround of two of India's most significant public sector undertakings—Hindustan Copper Ltd and National Mineral Development Corporation Ltd. I have had the pleasure of meeting the author Mr Rana Som when he was Chairman of NMDC. It would be easy to assume that the book is a specifically Indian story, but in truth the lessons that come from it are universal. Government-owned enterprises the world over struggle with the fact that they operate in a highly constrained environment which can put them at a disadvantage to their private sector counterparts. The constraints are financial, political and cultural, and the resulting impacts on the organisation are often negative—demotivation and feelings of powerlessness among management and employees, stagnation, resistance to change, inability to innovate, ineffective decision-making and failure to thrive. Mr Som's book shows us that these limitations do not need to be so. The story of the turnaround of Hindustan Copper Ltd and National Mineral Development Corporation Ltd contains important lessons that are relevant to all organisations, in particular the need to honestly evaluate strengths and weaknesses and to adopt strategies which leverage the strengths and deal with the weaknesses. Mr Som's book also reminds us that in any organisation, whether public or privately owned, change is ultimately driven by individuals. In government enterprises, it is not always easy for individuals to promote change from within, but those who are resolute and fearless can succeed in doing so.

Peter Jarosek
Partner, Corrs Chambers Westgarth, Australia

The Sleeping Tigers is a book of the same genre as that of Lee Iacocca's *An Autobiography* narrating the revival of Chrysler. But Rana Som's book is a step ahead because the organisations transformed by him were public sector giants in mining industry in India while Chrysler was a leading private sector automobile organization. The stories of the revivals are evidences of how important leadership is in any organization—be it private or public.

Raghabendra Chattopadhyay
Formerly, Professor of Public Policy and Management, IIM Calcutta

The Sleeping Tigers

The Sleeping Tigers
A Revival Story

Rana Som

Los Angeles I London I New Delhi
Singapore I Washington DC I Melbourne

Copyright © Rana Som, 2017

All rights reserved. No part of this book may be reproduced or utilized in any form or by any means, electronic or mechanical, including photocopying, recording, or by any information storage or retrieval system, without permission in writing from the publisher.

First published in 2017 by

SAGE Publications India Pvt Ltd
B1/I-1 Mohan Cooperative Industrial Area
Mathura Road, New Delhi 110 044, India
www.sagepub.in

SAGE Publications Inc
2455 Teller Road
Thousand Oaks, California 91320, USA

SAGE Publications Ltd
1 Oliver's Yard, 55 City Road
London EC1Y 1SP, United Kingdom

SAGE Publications Asia-Pacific Pte Ltd
3 Church Street
#10-04 Samsung Hub
Singapore 049483

Published by Vivek Mehra for SAGE Publications India Pvt Ltd, typeset in 11/14 pt Georgia by Fidus Design Pvt. Ltd., Chandigarh and printed at Saurabh Printers Pvt Ltd, Greater Noida.

Library of Congress Cataloging-in-Publication Data
Name: Som, Rana, author.
Title: The sleeping tigers : a revival story / Rana Som.
Description: Thousand Oaks, California : SAGE Publications, 2017.
Identifiers: LCCN 2017023368 (print) | LCCN 2017036893 (ebook) | ISBN 9789386602114 (Web PDF) | ISBN 9789386602121 (E pub 2.0) | ISBN 9789386602107 (print pb))
Subjects: LCSH: Mineral industries—India—Management—Case studies. | Corporate turnarounds—India—Case studies. | Industrial management—India—Case studies.
Classification: LCC HD9506.I42 (ebook) | LCC HD9506.I42 S66 2017 (print) | DDC 338.7/6220954—dc23
LC record available at https://lccn.loc.gov/2017023368

ISBN: 978-93-866-0210-7 (PB)

SAGE Team: Manisha Mathews, Apoorva Mathur, Shobana Paul and Ritu Chopra

*Dedicated to my parents
who taught me how to love the world.*

Rana Som

Thank you for choosing a SAGE product!
If you have any comment, observation or feedback,
I would like to personally hear from you.

Please write to me at **contactceo@sagepub.in**

Vivek Mehra, Managing Director and CEO, SAGE India.

Bulk Sales

SAGE India offers special discounts
for purchase of books in bulk.
We also make available special imprints
and excerpts from our books on demand.

For orders and enquiries, write to us at

Marketing Department
SAGE Publications India Pvt Ltd
B1/I-1, Mohan Cooperative Industrial Area
Mathura Road, Post Bag 7
New Delhi 110044, India

E-mail us at **marketing@sagepub.in**

Get to know more about SAGE

Be invited to SAGE events, get on our mailing list.
Write today to **marketing@sagepub.in**

This book is also available as an e-book.

Contents

List of Abbreviations	ix
Foreword by Dr Purnendu Chatterjee	xi
Preface	xiii
Acknowledgements	xix

Part I: The Story of Hindustan Copper's Revival   1

Chapter 1: The Task Assigned	3
Chapter 2: The Years that Preceded	10
Chapter 3: Opportunities Unfold	25
Chapter 4: The Decline	34
Chapter 5: Action Begins	42
Chapter 6: New Strategy	54
Chapter 7: Settling In	61
Chapter 8: Struggle Continues	67
Chapter 9: Process of Stabilisation	84
Chapter 10: Lessons Learnt	94

Part II: NMDC—The Sleeping Company Turns into a Giant   99

Chapter 1: A Great Company	101
Chapter 2: The Journey Begins	109
Chapter 3: NMDC Expands into Steel Making	119
Chapter 4: When Marketing Became the Key	154
Chapter 5: Struggle to Expand	171
Chapter 6: Managing People	193
Chapter 7: Community Work	206
Chapter 8: Lessons to Remember	219

List of Abbreviations

AITUC	All India Trade Union Congress
BALCO	Bharat Aluminium Company Ltd.
BHJ	banded hematite jasper
BHQ	banded hematite quartzite
BMS	Bharatiya Mazdoor Sangh
CCR	continuous cast copper rod
CEO	Chief Executive Officer
CITU	Centre of Indian Trade Unions
CLC	Chief Labour Commissioner
CMD	Chairman-cum-Managing-Director
CODELCO	Corporación Nacional del Cobre de Chile
CPI	Communist Party of India
CPSEs	central public sector enterprises
CSR	corporate social responsibility
CVC	Central Vigilance Commission
DAV School	Dayanand Anglo-Vedic School
EBITDA	earnings before interest, taxes, depreciation and amortisation
EC	environmental clearance
FOB	free on board
GSI	Geological Survey of India
HCL	Hindustan Copper Limited
HPCL	Hindustan Petroleum Corporation Ltd.
INTUC	Indian National Trade Union Congress
IRR	internal rate of return
KIOCL	Kudremukh Iron Ore Company Ltd
LC	letters of credit
LKAB	Luossavaara-Kiirunavaara Aktiebolag

LME	London Metal Exchange
MBA	masters of business administration
MCC	Maoist Communist Centre
MIC	metal in concentrates
MMDR Act	Mines and Minerals (Development and Regulation) Act
MoU	memorandum of understanding
MT	million tonne
NGO	non-governmental organisations
NMDC	National Mineral Development Corporation Ltd
PDC	post-dated cheque
PMO	Prime Minister's Office
PSU	public sector unit
RINL	Rashtriya Ispat Nigam Limited
SAIL	Steel Authority of India Limited
UNIDO	United Nations Industrial Development Organization
USSR	Union of Soviet Socialist Republics
VRS	Voluntary Retirement Scheme

Foreword

It is indeed a matter of great pleasure and honour for me to write this foreword. I have known Mr Rana Som, the author of the book *The Sleeping Tigers* for over 50 years now. After completing his studies, Rana chose to make his career in the Indian public sector and eventually rose to great heights. In the last 10 years of his illustrious career, he was chosen to lead two important central public sector enterprises in succession—the two 'sleeping tigers'. The first, Hindustan Copper Limited (HCL) was in dire financial crisis when he took over its reins. He successfully turned it around and made it a valuable company. The second, the National Mineral Development Corporation Ltd (NMDC), though profitable, was operating much below its potential. Rana transformed the company to a globally known mining giant.

Any turnaround or transformation of an enterprise—be it in the public sector or private space—presents a set of typical challenges to the leader. Manpower much in excess of requirement, uncompetitive and/or unreliable operating processes, lack of innovation or technology upgrade, inability to scale up to a premium specialised segment from a highly competitive commodity market, inability to position the products properly for securing right or superior value etc. More often than not, financial constraints tend to amplify the challenges, rendering these apparently insurmountable. Lucky are the companies that get the right leader at such critical junctures. A right leader is someone capable of viewing the issues holistically, understanding the intricacies and arriving at the package of solution that is likely to carry the buy in from most stakeholders. In the next step, the leader drives, rather relentlessly, the implementation of the solution package. The outcome is often seen to come in piecemeal, reinforcing each time the trust of the stakeholders on the leader. The journey finally ends in a celebration of the collective effort of every involved individual and with all round applause for the leader.

Reading the book, I could perceive the inner satisfaction that Rana rightfully derives from leading the process of turnaround of HCL and the transformation of NMDC. In both cases, he faced interesting challenges, though of different types. His ability to take the bull by the horn and lead the organisation from the front made all the difference. His anecdotal style of writing sustains the interest of the reader till the end. Understandably, even after the passage of considerable time, he fondly recollects the sequence of events vividly. It goes to show the passion, commitment and dedication that he brought to the assignments entrusted to him by the government.

I am sure that the book will find readership among teachers, students and researchers in management institutes. It can also become a valuable reference book for other public sector units (PSUs) and private companies striving to turnaround or transform.

Dr Purnendu Chatterjee
Founder and Chairman,
The Chatterjee Group (TCG)
Chairman, Haldia Petrochemicals Limited
Executive Director,
Indian School of Business, Hyderabad

Preface

Public sector enterprises played a very important role in India's journey towards building a modern nation from what it was left with at the end of the long years of British rule. Starting from 1948, when the first public sector enterprise, namely Indian Telephone Industries (ITI Limited) was set up in Bangalore, the number of central public sector enterprises kept on increasing over the years, and now there are more than 250 central public sector enterprises (CPSEs), including those which had been taken over from their private owners due to sickness.

Even before India decided to adopt the model of mixed economy with the public sector playing a leading role, there was a general consensus that an independent India should strive to secure economic growth through direct intervention of the state. In 1944, a group of leading industrialists issued what they called 'a plan of economic development for India', which conceded that the 'existing economic organisation based on private enterprise and ownership, has failed to bring about a satisfactory distribution of the national income' and 'enlargement of the positive as well as preventive functions of the state is essential to any large scale economic planning'.

India's industrial base before partition had not been significant. After independence, both the economy and the society were shaken by the tremors of partition. Political stability continued to elude the nation for some more years as the process of amalgamation with more than five hundred princely states took time to complete. In some cases, this involved armed intervention. The country did not have the foundation to encourage investments to flow in. Public sector took upon itself the task of building infrastructure, producing capital goods and bringing investments in the economy. In terms of volume, public sector in India is a mammoth representing an equity investment of ₹2,000 billion, a total investment and net profit of more than ₹11,000 billion each, employing 1.2 million people. To a large extent, Pandit Jawaharlal Nehru was right in describing the public sector enterprises as the temples of modern India.

However, the gross picture of Indian public sector tends to conceal its inner weakness. While 162 CPSEs are making profit, exactly half of that number are incurring losses. Many of the profit-making CPSEs are also not in a position to optimally utilise their assets. Public sector enterprises, barring a few, are known for sub-optimal performance, leaving huge scope for improvement.

Like any living organism, possibly, an organisation does also have a life cycle. An organisation's birth is generally followed by its growth and maturity and, thereafter, a process of stagnation and decline sets in. Most of the Indian public sector companies had not been any exception from this general trend. From mid-1960s, many public sector companies in India started showing symptoms of fatigue. Many of them failed to adjust themselves with the changing technologies and market. Many of public sector companies became hunting grounds for aspiring trade unionists and politicians. Weak management, poor work culture, lack of investable resources and misguided policies gave rise to a situation where many public sector companies silently surrendered to their fate and embraced sickness and closure.

It can easily be said that many organisations which are privately owned suffer from the same symptoms. Just like machines need timely maintenance and renovation, as human beings need periodical rejuvenation, an organisation also needs to be revamped and refurbished from time to time. If the fundamental health of an organisation is good, the job of its renewal is easy. The advantage in the public sector had been that most of these enterprises had been built with a lot of care. Most of them were constructed according to liberally prepared budgets without much compromise on cost and quality. The assets created in the public sector had, therefore, generally been of high standard. What the public sector management lacked was the understanding that these assets had to be sweated for extraction of value which would justify the cost of the asset. As long as the process was duly supported by the technology and the product was duly accepted by the market due to monopoly, overall scarcity or otherwise, public sector companies did not have any problem. The problem started when the resources became scarce due to poor internal generation of funds and replacement of machines or the adoption of new technology became difficult. The problem was aggravated when the market demanded new products requiring new investment or new initiative or offered lower value for the existing products. In industry, it is an inevitable truth that profit is nothing but a short-term benefit of an innovation as Joseph Schumpeter, the renowned American economist once stated. With time, the products are copied and the technology is imitated. Therefore, as the time progresses, one has to take initiative not only to produce but also to upgrade. This did not happen in the public sector and that was the problem not only with public sector companies but also with those in private sector who did not follow the golden rule of setting aside funds for renewal.

When an analysis is made on this issue, as to why this was not done, it appears that in public sector, the failure was because of the short tenure of the management. A manager in a public sector, especially those on the top, had once had very brief tenure of three years, and even now the average tenure is not more than five years. This prevented the managers from taking a long-term view. Second, the militant unions wanted to extract maximum benefit from the public sector companies not only by way of higher wages but also by pushing unproductive people into the company. Lastly, the owners of the company, that is, the state, did not appreciate and understand the long-term requirement because the state itself was represented by a continually changing group of people, both bureaucrats and politicians. In a privately owned company, similar situations come up mainly because of the extremely short-term view adopted by the owners who prefer to keep whipping the old horse rather than replacing it with a new one.

In *The Sleeping Tigers*, stories of two large public sector enterprises have been told, one of which was perennially sick. For seven-and-a-half years, the company was losing money, and at one stage the loss exceeded ₹100 million per month. The accumulated loss of the company exceeded ₹7 billion. The situation was managed by some simple methods. Use cash credit limit in full, get support from the government, defer salaries and wages, withdraw benefits and use imported material under a letter of credit (LC) and use it for meeting short-term liquidity needs. These methods can defer the problems, but cannot solve them. The obvious consequence had been that the situation came to a breaking point when there was no option on the part of the state but to simply abandon the company by way of strategic sale. The other company, NMDC, had strong fundamentals, a good market and good production. But the company remained in a static position for years. Nature denounces stagnation: whatever is stagnant is set to decline. In this company, the symptoms of decline had just started appearing when the interventions were urgently required.

Both the companies had inner strengths, but the people were either demoralised or complacent. They belonged to strong organisations, but because of their inaction, the organisations looked helpless like sleeping tigers. *The Sleeping Tigers* is a story of these organisations and how they were turned around through the effort of the same people who had almost surrendered and left the company to its fate.

When this book was written, the intention was not to draw credit for the successful turnaround of the concerned companies; the purpose was not even to highlight the achievements and publicise them. The purpose was only to share the challenges, strategies, actions and emotional involvement of the people at every level that constituted the process of the turnaround. It was an exercise for the existing and future managers to learn from. It is an effort to generate hope in all those who feel that public sector companies in particular do not have any future, and if they are sick, they should be allowed to die. The book shows that revival and renewal of a public sector company with strong assets is much cheaper and easier than creating new entities. It shows that people's motivation is the key factor in changing the fortune of a company. It shows how creativity is an inherent quality and how it can be nurtured in a right environment. For any process of renewal and revival, there is nothing more important than the human factor, as it is highlighted again and again that organisation means nothing but a group of people who work for it.

Another factor which is very important in a turnaround process is the synchronisation of various activities and coordination among various functions within the organisation. A good organisation works like a symphony. Its rhythm breaks even if a small instrument fails to function. The work of an organisation in such an integrated manner requires human motivation, skill and alertness on the part of leadership. It is a very demanding but rewarding process. Success in turnaround comes to a large extent from such coordinated and holistic efforts. The book depicts how work was carried out simultaneously in different fronts and whatever work was done was communicated to everybody in the organisation. This approach had produced great value, both in terms of employee participation as well as in achieving results.

Lastly, what the book evidently shows is that there is no substitute to hard work and obviously placing the organisation over self. There is no miracle in a miraculous turnaround. For rebuilding an organisation, every brick needs to be laid one by one, every structure needs to be measured properly before installation and every move needs to be taken with caution. When the organisation picks up momentum, people develop the habit of better performance, better productivity and better cooperation. They learn to take pride in their organisation, and this pride keeps the organisation moving. This is what *The Sleeping Tigers* lets us learn and believe in—it is a story of faith and renewal.

Acknowledgements

The Sleeping Tigers is not a fictional tale. It is a book which intends to record the spirited contribution, determination and dedication of a large number of working people to rediscover and re-establish their internal strength, bring their organisation out of sickness and stagnation and put them firmly on the path of revival and rejuvenation. With all humility, I acknowledge their courage to think and do something different, but for which *The Sleeping Tigers* would not have been a story worth telling.

I also acknowledge the contribution of a large number of people including politicians, bureaucrats, bankers, journalists and even small suppliers who did not belong to the organisations concerned but consistently supported the process of turnaround, encouraged it and shared the happiness of the employees at each turn of events that took the organisations forward. Many of them were prepared to take risks and often went out of their mandate to firmly stand by the people who were struggling with a smile.

The book is a direct product of the inspiration I received from my wife whose persistent persuasion did not allow me to escape from this task. My secretaries happily worked extra hours to take dictations, do typing and edit the text with so much dedication that any words to thank them will be inadequate.

My sincere thanks to SAGE Publications for undertaking the responsibility to print and publish the book.

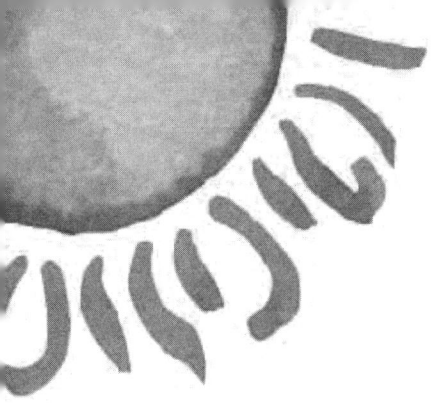

Part I

The Story of Hindustan Copper's Revival

1

THE TASK ASSIGNED

It was 29 November 2003; the brief winter in Kolkata had just begun. The weather was excellent, free from heat and humidity, but the atmosphere within the company was not as comfortable as the environment outside. Hindustan Copper Ltd (HCL), one of the large public enterprises in the mining sector, came to the brink of closure. The Chairman-cum-Managing-Director (CMD) of the company was to retire on 30 November, but no initiative was taken by the government to find his replacement, presumably because it had been evident by then that the company, which had been the only copper mining company and custodian of all known copper deposits in India, was going to be closed down as a result of extreme financial crisis with no hope in sight for the company's business to revive. On the last day of the week, that is, on 28 November, the Chairman was granted an ad hoc extension for three months by the Honorable Minister concerned, which was to go later to the appointment committee of the cabinet for post-facto approval. Apparently, the only job left to be carried out by the outgoing CMD was to put the company under care and maintenance latest from the last week of December 2003, beyond which the company could not be run. The letters of credit (LCs) issued by the bank for import of copper concentrates had already been extended by more than 500 days and there was no way to get further extension. Freezing of the company's bank accounts was imminent. Some other liabilities including payment of power bills were equally pressing. Collection from sales during the month of December was estimated to be less than one-third of the minimum and emergent liabilities to be met. Salaries were already delayed by three months. Most of the mines were closed. The operation of the smelters was bleeding the company further as even the variable cost could not be recovered. There was an air of uncertainty all around. There were talks of shifting the headquarters of the company from Kolkata to Delhi and to sell off its corporate office building in Kolkata, for which there was a ready buyer. The employees with the old pay scale were entitled only to a meagre salary. The only option left with them was to seek premature retirement under a voluntary retirement scheme and settle for a small amount as compensation.

Just a few months back, the effort on the part of the Government of India to divest the company through strategic sale had failed. The government was reportedly ready to give the buyer the maximum possible comfort to close the transaction in the form of payment of a meagre amount which could have been as low as ₹100 million. However, the deal could not be concluded, not because of dire financial position of the company but due to a technical hitch. The Ghatsila unit of Hindustan Copper including its mines and smelter earlier belonged to the Indian Copper Corporation, a privately owned company. The company was nationalised and merged with Hindustan Copper under an Act of Parliament and, therefore, as per the rulings of the court in a similar case, Hindustan Copper along with the merged entity could not be sold without the approval of the Parliament, which was difficult to secure. There was a thought to demerge the units of Indian Copper

Corporation and to close them down separately in order to make the rest of the company free to be divested, but this too did not work out as it was also found to be legally infeasible. The trade unions had vociferously agitated against the proposal to disinvest the company, but then, when the proposal for disinvestment was dropped, the situation did not bring any solace to them. Unions realised that the only other option was the closure of the company and its gradual extinction. By that time, I had been in the company for three years, I had a fair assessment of the inner strengths of the company and, therefore, I did not lose hope. Hindustan Copper never deserved to be written off. During my long years of service, I had observed that a company which possesses natural resources can be better off or worse off, depending on the market situation. But its fundamental strength lies in the resources it has. A mining company, therefore, cannot be compared with a purely manufacturing company, where change in process, technology, cost or competition can make a company incurably sick. A company having natural resources needs to be protected by the state or the society because closure of the company can amount to a loss of resources, not only for the company but also for the nation at large. Therefore, a mining company stands altogether on a different footing as compared to other types of companies. Hindustan Copper's biggest strength had been that it was the only copper mining company in India. India was never known for its copper deposits. In fact, just about one per cent of global copper reserves were in India, but whatever reserves of copper the country had were being mined by Hindustan Copper, and that was the primary strength of the company. As long as copper as a metal did have value, Hindustan Copper remained valuable. Yes, the company was not doing well financially because of a plethora of reasons, but then many of them could be rectified through managerial intervention. I felt that Hindustan Copper was in a way like a sleeping tiger which looks helpless as long as it is asleep, but if the company could be turned around and the tiger gets up from its deep slumber, it would be back to its glory and strength. I was keen to put in all my imagination, ideas, strength and passion towards reviving the company.

The opportunity came on Saturday, 29 November 2003. I was in the office as usual on that day. At about 11.00 AM, I received a call from the administrative ministry. A senior officer asked me where I was. Apparently, he did not know that I was in the office. He told me that he was also in the office, which was a little unusual because officers in government departments in Delhi were very rarely in office on Saturdays. The gentleman advised me not to leave the office until I got a message from him. I thought that it would be impolite to ask him what the message was about. I got another call at 1.30 PM from the same gentleman who reminded me once again that I should not leave the office. I was a little surprised and could not resist the temptation of asking him what this was for. He said that something was happening in the Cabinet Secretariat (which too was working on that day) on some issues relating to Hindustan Copper. I waited in anticipation. For Hindustan Copper, any news should have been good news because there could not be any situation worse than what the company was going through.

Once again, I got a call at about 3.00 PM. This time, the gentleman said that the appointment committee of cabinet had considered the provisional approval of the Honorable Minister to give extension to the existing CMD and decided to reject it. It had further been decided to appoint me as the officiating CMD. I was advised to continue in the post until it was formally filled. It was a surprise. I had been working as director (personnel) in the company for exactly three years by then. I had joined the company on 30 November 2000, as its director (personnel), and it so happened that on 29 November 2003, I had just completed three years as a director on the board. By then, two other functional directors had left the company on retirement or resignation. I was the only director left in the company to be given the charge. Apparently, it was not a great deal, but then it was a great responsibility. I was trying to assess whether it was an opportunity or a threat. My inner soul gave me an answer within a few seconds. It was neither an opportunity nor a threat. It was a challenge which comes to a person once in a lifetime, and one must embrace it with humility, courage and sincerity. While working in the company for the previous three years as a director, I developed my own opinions on how the company could handle the crises which were slowly engulfing the company. My views and suggestions had also been communicated to all concerned before. However, these had either not been properly heard or not been appreciated. I found that now there was nobody to stop me from doing whatever I felt was right. I realised what thousands of desolate eyes were asking for. They were ready to make any sacrifice. After all, by then they were left with only one animal instinct and that was to survive.

The company's financial condition was so precarious that in the month of November 2003 itself, the company had incurred a loss of ₹100 million, further adding to the already accumulated loss of about ₹7 billion. Most of the marketing offices of the company had been shut down sometime back, because the then CMD decided to reduce costs by giving early retirement to the officers and staff in those offices. The company's production from mines was at an all-time low. The smelter in Khetri in Rajasthan was running with imported copper concentrates. The process was uneconomic. The copper concentrates were imported through Kandla and transported all the way to Khetri. After smelting and refining, copper cathodes produced in Khetri were again transported to Taloja in Maharashtra to be converted into wire rods and transported once again to customers all over the country. The cost of logistics was high and so was the cost of smelting and refining in the old smelters in Khetri and Ghatsila. Hindustan Copper was losing on every tonne of copper cathodes or wire rods it produced. Despite this, the company continued to import copper concentrates for the apparent reason that it helped the company in its cash flow management. Import through LCs did not involve immediate cash outflow, but the sale of finished copper did provide cash in advance. Hence, it was considered to be a solution for the cash flow problem of the company. In the ultimate analysis, however, it amounted to nothing but deferring today's problem to tomorrow and making tomorrow more difficult. The company's last LCs were extended by more than 500 days with the grace of a public sector bank, but the bank's patience had reached a breaking point. The bank had issued a final notice that their board had decided to seize the bank accounts of Hindustan Copper in December 2003 if there was any further default. There was hardly any money in the bank, but freezing of the bank account would make the entire company non-functional. A consortium of banks led by the State Bank of India had given Hindustan Copper a credit limit of ₹1.65 billion, but insisted for a government guarantee to support it. The Government of India had agreed in principle to provide some budgetary support to the company, but informed that this would not be available before the end of March 2004. In September 2003, the company was unable to make payment of electricity bills. Fuel supply had stopped because the payments were already overdue. The stock of **imported** raw material was nil and the internal production **of copper** concentrate was meagre because just a few months back the Khetri mines had been closed. With the Khetri mines closed, the other neighbouring mine in Kolihan became uneconomic because the overheads of Khetri were now loaded on Kolihan, making it unviable. Only two things kept me inspired. First, I was now responsible for about 6,000 people and their families and they looked helplessly towards me. They felt that they had been abandoned by the state which owned the company, that they had been ditched by the management which did not run the company properly and that they had been bulldozed by the company into accepting the closure of the mines one after another. Second, I was always convinced that the ideas which I had cherished were the right ideas and this was the time when I could implement them without hindrance. It was a testing time for the manager within me. It was an opportunity for me to elevate myself from the role of a manager to that of an entrepreneur. This is how my journey began. I turn to the next chapter by just saying that in the month of December 2003, Hindustan Copper was not shut down. The company was not put under care and maintenance. The company made a net profit of ₹7.7 million in December 2003 after continuous loss for 90 months and accumulated losses of more than ₹7 billion. A dying company fought back, and it fought back with all its strength, which it finally discovered that it did possess. After December 2003, Hindustan Copper never looked back. My story is the story of this turnaround; it is the story of the people's determination and of the revival of a company which almost everybody felt was sunk forever.

2

THE YEARS THAT PRECEDED

My journey in Hindustan Copper commenced on 30 November 2000. My professional specialisation was in human resource management, but then I always felt that management was too important an exercise to be fragmented into different segments. For a manager, the company has to be viewed as a whole and the approach of management has to be holistic. The ultimate job of a manager is to make a sustainable profit, not only profit for today but also for the days to come, and this can happen if along with company's business, all the stakeholders can be taken care of. The customer has to be happy with the quality and price of the product. The employee has to be happy with his/her job, growth and remuneration and recognition. The technology has to be adequate for the business. The environment has to be protected so that there is no conflict between the business and the environment. The role of every manager in the company is so interlinked to the total business that any short-sighted sectorial approach can be disastrous for a company. In fact, such sectoral or departmental approaches are often the main causes behind managerial failures in most organisations.

After joining Hindustan Copper, I first wanted to have an idea about the copper industry which I was never associated with in the past. I was also keen to understand the company, its operations, its infrastructure and its position in the industry at the national level.

Copper is one of the oldest metals man has discovered. Even 10,000 years ago, man knew about copper and its usage. A copper pendant discovered in northern Iraq has been dated to about 8700 BCE. The word copper comes from the Latin word of *cuprum*, meaning 'from the island of Cyprus'. Copper is the only metal other than gold that has natural colour. Other metals are either grey or white. The use of copper and its contributions continued to grow as human civilisation progressed. Over 400 copper alloys are in use today. Its inherent properties include its attractive appearance, high conductivity, good corrosion resistance, ability to alloy with other metals and ease of working. Copper has entrenched into the expanding markets in electrical, electronics and communications industries. It is also used for automobile radiators, cooling and refrigeration tubing, heat exchangers, small arms, ammunition, water pipes and jewellery. It is valued for its heat and electrical conductivities and malleability. The world was known to have copper reserves of more than 300 million tonnes in the late 1990s, which as per the latest figures has gone up to about 700 million tonnes in 2015. Chile has the highest reserves with 29.9 per cent followed by Australia with 13.2 per cent, Peru with 9.7 per cent, Mexico with 5.4 per cent and the USA with about 5 per cent of the global reserves.

In terms of production, Chile has always been the largest producer of copper now followed by China, Peru, the USA and Australia. Production of copper has been showing steady increase over the years. It has grown from less than half a million tonnes in 1900 to more than 13 million tonnes in 2000 and more than 16 million tonnes at present. Both in terms of reserves and production, India has always been a minor player. India's copper production constitutes just about 0.2 per cent of the world production. In 2000, when I joined Hindustan Copper, India was producing just around 35,000 tonnes of copper from its mines located at Khetri in Rajasthan, Malanjkhand in Madhya Pradesh and Ghatsila in Jharkhand. Even though, as compared to the global figures, India's copper production was insignificant, Hindustan Copper was the only copper mining company in India, except for Sikkim Mining Corporation which used to mine a very small quantity of copper. The entire production of copper in India was practically contributed by Hindustan Copper only. The company also operated two smelters and refineries producing copper cathodes, one of them located in Khetri and the other in Ghatsila.

Hindustan Copper was set up on 9 November 1967. The National Mineral Development Corporation Ltd (NMDC), another Government of India enterprise which I headed later, had constructed the underground mine in Khetri in Rajasthan and Rakha in Jharkhand for the extraction of copper ore. Hindustan Copper took over these mines from NMDC. The company became responsible for indigenous production of copper. It got engaged in identifying new copper deposits, developing them and producing pure copper through a process of beneficiation, smelting and refining. The process starts with the mining of ore. Copper ore contains only a small percentage of copper metal with the remainder of the ore being unwanted rock or gangue materials which does not have much commercial value. Usually, the metal content in copper ore varies from 0.6 per cent to 2 per cent. Therefore, the ore requires an elaborate metallurgical treatment for the separation of minerals from gangue materials within the rock. Beneficiation or concentration is the first stage of the metallurgical treatment which is done through a process called froth flotation that requires a large quantity of water. The beneficiated ore, which is called concentrates, contains around 30 per cent copper and is charged into smelters for the extraction of copper blisters with about 98 per cent purity in an anode furnace from which copper anodes are produced with 99 per cent purity. The anodes are then fed into the refinery where about four weeks' time is required to produce copper cathodes of 99.99 per cent purity. The material which settles down at the bottom of the cells is called anode slime. It contains by-products such as silver, gold, selenium and tellurium. Hindustan Copper's main products had been copper cathodes, which were produced in Khetri and Ghatsila, and copper wire rods, which were produced in Taloja in Maharashtra. The company's smelters in Khetri and Ghatsila had capacities of producing copper cathodes of 31,000 tonnes and 16,500 tonnes per annum (pa) respectively.

The most important asset of the company was the copper mining project in Malanjkhand in Madhya Pradesh. Unlike Khetri and Kolihan, which were underground mines, Malanjkhand had open cast operations. It had reserves of about 200 million tonnes of copper ore, which contributed more than 40 per cent of India's copper reserve. Incidentally, as stated earlier, India was never known to be rich in copper deposits. About one per cent of the world's copper reserve that was found in India was mainly available in three states, namely Madhya Pradesh, Jharkhand and Rajasthan. Malanjkhand mines had a capacity of producing 2 million tonnes of ore pa with 1.3 per cent copper content. The unit also had a concentrator plant for producing copper concentrates from ore by using the process of floatation.

The third unit of the company was located in Ghatsila in Jharkhand. This unit belonged to the Indian Copper Corporation Ltd, a private sector company which was nationalised in 1972 and made a part of Hindustan Copper. Apart from having a small smelter and refinery, the unit had a precious metal plant for extraction of gold and silver from copper slimes. Its by-products also included sulphuric acid, nickel sulphate, selenium etc.

The fourth unit of Hindustan Copper was located in Taloja, Maharashtra, near Mumbai which had a capacity for producing 60,000 tonnes of copper rods pa from its continuous cast copper rod (CCR) plant. The plant used copper cathodes brought from Khetri and Ghatsila for producing cast copper rods. This involved a long journey for the copper cathodes from one end of the country to another end and added to cost, putting further strain on the company's cost structure.

Hindustan Copper had already been incurring losses for the previous four years. Production was steadily declining. Ore production fell from about 4.5 million tonnes in 1997–98 to just about 3 million tonnes in 1999–2000. The production of copper cathodes declined from 42,000 tonnes to about 36,000 tonnes by using internal production of copper concentrates as well as imported concentrates. Lower production and lower capacity utilisation of plants and mines further worsened the situation. Meanwhile, two new copper smelters and refineries came up in the west coast of the country. One was at Dahej in Gujarat, which was set up in 1991 and had an initial capacity of producing 100,000 tonnes of copper cathodes pa, and the other one was at Silvassa which was set up by Sterlite Group. These smelters were shore-based and they were only engaged in the conversion of copper concentrates into copper cathodes. They had to spend least on logistics and took advantage of modern technology and economies of scale in their operations. Compared with these smelters, the smelters in Khetri and Ghatsila were small and old. The condition of the mines was also not up to the standard because of lack of investment for a long time.

During the first few weeks of my joining Hindustan Copper, I visited all the operational units one by one. I felt really proud. I went into the underground mines in Khetri and spent time to see and understand the operations of concentrator, smelter and refinery. All the facilities were compacted into one place for deriving the best operational economics. I went to Malanjkhand and saw the ease of operations that an open cast copper mine could provide. From Malanjkhand, copper concentrates were transported to Khetri or Ghatsila for smelting and refining. In Ghatsila, the residential area looked more attractive than the mines and smelters. The township in Mosabani, which was built by the Europeans in the past, even though shorn of much of its glory, looked like a leaf taken from a picture book on English counties. The golf course of Moubhandar was world class. Most of the mines in Mosabani were too deep to be mineable and were already closed. Only two mines in the area, namely Surda and Rakha, though producing small quantities, were functional. The precious metal plant for extraction of gold and silver from copper slimes, though not very efficient, was then producing more than 500 kg of gold and 7,000 kg of silver pa. The copper-producing company was also a large producer of gold and silver.

The assets of the company appeared huge, but the level of value extraction from those assets was poor. The company's employee strength was about 13,500, down from about 26,000 in the olden days. There was consensus in the company, which I discovered soon after my joining, that the manpower had already been downsized to such an extent that it was impossible to take care of all the assets properly with lesser number of people, and that it was absurd to think of reducing the manpower further. Unions were strong and their strengths and visions were so well-known that within a couple of days of my appointment, I was told that my first job would be to firmly handle the unions. It was genuinely believed that they were responsible for most of the evils the company was facing. Union militancy was clearly visible; so was inter-union rivalry.

There were plenty of restrictive practices; some were so alarming that it was difficult to believe that those could have perpetuated for such a long time. In one of the units, only 50 per cent workers would come for work in the beginning of the shift and go away during the recess. Another 50 per cent would come after the recess and remain till the end of the shift. Thus, in every shift there was double manning and each shift instead of eight hours became one of two shifts of four hours each. What surprised me was that everybody knew it, but preferred to ignore it lest the unions got angry. When this was discussed with the unions separately, each one of them said that they knew that it was bad and that they condemned it, but that it was not their job to maintain workers' attendance, and even if they supported the management in removing such practices, the competing unions would foil such an attempt. If the unions took advantage of a weak management, so did the police and state officials. Many of them forcibly occupied the company quarters and enjoyed the facilities. The company looked like a harassed entity with poor performance, financial strain, external invasion and internal rot.

My first impression of pride soon turned into agony and anger. The situation gave me an impression that prolonged sickness of the company had given a sense of hopelessness and despair at every level, including, possibly, the highest level of the management. Talks of disinvestment had already begun. Workers were extremely apprehensive about the future. Those at the top management knew that they would have to give way to a new set of people to take over management of the company.

> I had only one dilemma whether inadequate management, indiscipline and aggressive unionism were the cause of the situation the company had landed up into or whether those were the results.

To me, it appeared that both were partially correct. The cause of sickness was inherent in more critical factors, namely low London Metal Exchange (LME) price of copper, old technology, very high manpower and emergence of new players in the copper market who had shore-based plants with high capacity, latest technology and advantage of economies of scale in their favour. But then, low morale of employees and management failures had aggravated the problem to a large extent. Meanwhile the market share of Hindustan Copper had come down to about 6 per cent to 7 per cent in the Indian copper market. The situation was completely unenviable, and no scope of recovery was apparently visible.

I started my bit of work immediately. I picked up the attendance issue to handle first. I decided to meet the union leaders one by one on a single day to sort out the issue. I knew that the leaders had hardly any defence, but exposing their follies or defeating them in arguments was not the solution. I, as a new entrant in the company, sought their support in removing a practice which was indefensible. Each union leader said that they did not have any problem in supporting the management to stop the wrong practice if others co-operated. I moved to another leader with the same request and got the same answer. In the end, I assured the union leaders that if they agreed, others would also agree and each one separately agreed. Next day we issued a notice to all employees advising them to be punctual and explaining how important that was for the company. When the workers approached their respective leaders, each leader said that they were helpless because leaders of other unions had agreed to stop the old practice and in any case it was difficult to defend that. There were a few stray incidents, but overall the situation came under control. Within a few days, the new order became an accepted system. A practice that had lasted for years ended so quickly and so smoothly that I was forced to wonder why it had lasted for such a long time in the first place. In one go, we could identify the extent of surplus people the company was carrying, and with that the myth that the company did not have further scope to rationalise manpower further was busted.

> When I think of the incident in retrospect, I feel that one of the reasons why the wrong system continued for such a long time was because not only the workers and their unions but also the management tacitly accepted the position and refrained from acting against it.

It is always easy to blame the unions and escape into inaction than to catch the bull by its horns. The unions on the other hand did not gain from the system, but they felt that they would lose if they took any initiative to change it, especially when inter-union rivalry was strong. Workers knew that what they were indulging in was unjustified but enjoyed the helplessness of both the management and the unions. I had observed similar situations in other public sector undertakings (PSUs) as well where the management was controlled by unions and unions by workers. In such circumstances, unions actually need management action to come out of unfair control of the workers so that they can regain the ability to lead. The attendance issue and my intervention helped the unions to come out of the clutches of the workers who were pursuing wrong action.

An important development occurred within the next couple of weeks. The unions jointly served a strike notice on the management. They demanded revision of wages, which was already due for about four years, and insisted that until the wages were revised they would have to be paid an interim relief at a certain percentage of pay. The situation was tense. After a few days, the Secretary of the administrative ministry came down to Kolkata and held a meeting with the Chairman and Directors. When the issue of strike notice came up, he expressed his surprise. He wondered how any responsible union could serve a strike notice at a time when the company was just struggling for its survival. The CMD and other directors agreed that the demand was absurd, unreasonable and unacceptable. The matter did not deserve even a discussion. As a new entrant into a team, I was the last person to get an opportunity to speak. I told those present in the meeting that whenever a strike notice was given, the unions knew that negotiations were inevitable. They also knew that if they had to get something, they were also to give something back, and that was the principle of collective bargaining, which was nothing but an art of exchange. The demands for wage revision and interim relief were close to the heart of every worker and by serving a strike notice on those demands, the unions had ignited the expectations of the workers. They will now be more ready to offer more co-operation in exchange of what they expect from the management. Hindustan Copper should use this opportunity to come to a win-win solution. I explained that the job of the management was to convert a threat into an opportunity and this was an ideal situation where it could be tried. The company was passing through bad days and it had strong unions. The company could sort out many ongoing issues with the active help of the unions. If whatever the unions conceded was more valuable than whatever they demanded, the company should not have any problem in acceding to what they were asking for. I further explained that the company, despite some past rationalisation of manpower, was carrying excess manpower in every sector. The number of employees could be immediately cut down by at least 2,000 if not more. The age of superannuation was already planned to be reduced by two years, though it could not be implemented. If done, then it would facilitate further reduction of manpower through natural attrition in addition to an effective voluntary retirement scheme. If this could be achieved, then Hindustan Copper would come into a position of acceding to the unions' demand. My colleagues and superiors present in the meeting were surprised and could not refute the logic in any way. I stated that the negotiations would be slow by design and that we should expect results only after a few months.

The negotiations started at the level of Chief Labour Commissioner (CLC), New Delhi. Hindustan Copper was a public utility service and therefore, the matter was seized in conciliation at the level of CLC. I was representing the company.

During the first meeting, I kept on telling the unions and the CLC why it was impossible to even consider the union's demand. I explained the financial position of the company and alleged that unions were never concerned about the health of the company and instead of assisting the company, co-operating with it and suggesting measures for improvement, unions always played their conventional role of making demands and fighting for their realisation. As expected, this provoked the unions adequately. The unions blamed the management for mismanagement, thereby giving me further opportunity to go into a more detailed discussion about everything that concerned the company and its employees. As soon as this happened, the negotiation got converted into a session of understanding each other's views. The unions kept on pressurising for some commitments, none of which was given. At about 11 that night, the CLC, who was known to be an efficient officer, requested for a break. He requested both the parties to resume discussions at a bilateral level and to jointly find out ways and means to reach a negotiated settlement. When we came out of Shram Shakti Bhavan, the roads were deserted. All of us were tired, and for the unions, nothing was achieved. But then, there was no bitterness. The unions were satisfied that at least they were heard with patience and sincerity. They were convinced that ultimately logic would prevail. I was satisfied because the real platform for moving forward was already built.

The next meeting with the unions took place after a gap of more than one month. This meeting lasted for three days. We came out with a power point presentation. Every allegation made by the unions was given a reply. Every mistake committed by the management was admitted. The unions ultimately found themselves in a position where they had to decide whether their demands were possible to be met or not. They had made their demands without knowing their financial implications. Now they were given the task to find a solution from all the facts presented to them. The unions were already convinced that the company could not be run in the same way as had been done during the previous years. They now knew that there was surplus of people and that, as in many other sick companies, the luxury of having 60 years as age of superannuation could not be afforded in Hindustan Copper. In fact, until 1998, the age of superannuation was 58 years. The company presented its proposal to roll back the age of retirement to 58 years as it would create a situation wherein a good number of employees could retire voluntarily under a scheme of voluntary retirement which was funded by the government. After the new system of attendance was introduced, existence of surplus work force was visible. People, especially those in Khetri, could clearly see that their plant needed just about 50 per cent people as compared to the number deployed earlier. I assured the unions that the voluntary retirement scheme would remain voluntary and there would not be any coercion either on the part of the management or unions to force the employees to opt for voluntary retirement. The unions were assured that their demands for interim relief would be very favourably considered after shedding off the excess manpower. Both the unions and the management worked together to make some marginal improvement in the voluntary retirement scheme to make it more attractive and more humane. The unions, after prolonged discussions among themselves, came out to support the proposal of reduction of the age of superannuation from 60 years to 58 years. They were not only internally convinced; they also had to meet people's expectations, which they themselves had raised. The work force in Hindustan Copper was fully unionised and the unions, having placed certain demands which had widespread support from the workers, could not let their members down. The unions not only agreed for revision of the age of retirement but they also committed that they would co-operate with the management in manpower reduction through voluntary separation. When the voluntary retirement scheme was reopened, 2,400 employees opted for voluntary retirement. The savings in salary and wage bill were four times the cost of interim relief. The company without delay announced a grant for interim relief with arrears for a few months. While making the demand for interim relief, the union leaders were themselves not convinced that their demands could be realised. For them, the call for strike was a little adventurous, but they had to give the call on that demand because they were under pressure from their supporters which they could not resist.

When the age of superannuation was revised as 58 years, it was applicable not only to the workers and other employees but also to the CMD and directors. However, not only Hindustan Copper but some other sick companies were also rolling back the age of superannuation to 58 years. The matter was taken up by the union cabinet for a review and it was decided that whenever a company rolls back its age of superannuation, it would apply not only to the workers and the officers who were appointed by the company but also to the board-level functionaries, who were appointed by the President of India. Therefore, the rolled back age for superannuation was effective for everybody right from the Chairman of the company to an employee of the lowest rung.

The Chairman of Hindustan Copper, who was to retire in 2005, straightaway lost two years of service, which he really did never expect to happen. It was a turning point, not only for him but also for others, including the company as a whole.

SOME INTERESTING FACTS

- When the company issued notice for voluntary retirement, giving four weeks' time for employees to apply, the company received only a few applications till the end of the third week. The unions supported the scheme, but had remained mostly passive till that time. The leader of the largest union was then contacted. It was once again explained to him how important his active support was for the scheme to succeed as the company's survival depended to a large extent on reducing manpower. He heard, but did not respond. After three days, I came to know that the person concerned who had many years to go at that time had himself submitted his application for voluntary retirement. When I called him up, he said that there was no other way left. Unless he himself applied for voluntary retirement, how could he tell others to do so. During the next three days, more than 2,000 applications poured in. Ultimately, the number of employees who took voluntary separation exceeded 2,400.
- An outcome of the long negotiations held with the unions had been that both sides got immense education about the company and the aspirations of the employees. Both management and unions learnt how to develop empathy for each other's position. This had a long-lasting impact. Years later, when I was no longer a part of the company, two senior leaders telephoned me and explained the structure of pay revision which they were expecting from the management. Their question

was whether what they had demanded and expected the management to agree to was right from the point of view of company's overall health. If not, they would scale down their demands to whichever extent I would suggest. I had no words to express my appreciation for such a laudable gesture and their positive attitude.

- During the whole process of negotiation, the role of the Government of India was exemplary. The Secretary of the ministry not only appreciated the strategy but he developed a trust that the management team would be able to implement the plan to achieve a win-win solution for everybody. I enjoyed his trust and confidence, even at a personal level. However, all the top executives of the ministry were not on the same page. When the voluntary retirement scheme was introduced, the time given for the operation of the scheme was 30 days and the internal target for securing applications for voluntary separation was fixed as 2,000. When 15 days had passed and only 100 applications were received, a very senior officer of the ministry declared the scheme a failure because in his calculation, in 15 days on pro rata basis, at least 1,000 applications should have come and, therefore, the percentage of failure was 90 per cent. Ultimately, more than 2,400 applications were received and the company succeeded in not only achieving but also surpassing the target despite such cynicism and sarcasm which fortunately was not shared by many.

3

OPPORTUNITIES UNFOLD

My boss in Hindustan Copper, the then CMD, had many qualities, but he was never known for the quality of being able to hide his inner feelings. After the rollback of the age of superannuation in the company, he was annoyed, and he had reasons to be so. He just could not tolerate me. One afternoon, when Kolkata was burning in the mid-summer heat, he called me and said 'you go to Khetri immediately'. There was no problem in Khetri of which I was aware. I was a little anxious and asked him what happened and what was I expected to do there. He said, 'I do not know, you just go and be there. I do not think you have got anything to do here'. I was shocked, pained and felt humiliated. I struggled to smile and said whatever he said would be done. Next morning I flew to Delhi and took a vehicle to Khetri. If Kolkata was a frying pan, Khetri was fire. But I consoled myself thinking that it was the desire of God and my job was to do my best in the given circumstances and try to convert a threat into an opportunity. Incidentally, this trip gave me the scope to develop a deep insight into the operations of the company, understand the basic problems and find out which part of the problems could be handled through managerial intervention and which were beyond management control. I knew that the global price of copper would not improve through our effort, but I also knew that the operations could be improved by cutting down cost, improving efficiency, motivating people, restructuring deployment and upgrading technology to the extent possible, within the resources available with the company. I knew that a passionate involvement of someone in the top management could be useful to bring in changes in a positive direction.

I had been to Khetri twice before, but those were short visits either to get an introduction of the project or to attend formal meetings. This time I went to Khetri under different circumstances. I decided to dive deep into the critical issues after understanding the project in a holistic manner. It also gave me an opportunity to learn more about the intricacies of copper mining and metallurgical processing of copper ore. I learnt about the area, its past, its topography, its people and the operational metrics of the entire project, which comprised two mines, a beneficiation plant or concentrator, a smelter and a refinery. After a few years, when I went over to NMDC as its Chairman, I never hesitated to admit that, in terms of effort and operational complexities, iron ore mining was nowhere near mining and mineral processing of non-ferrous metals such as copper, gold, silver etc.

Khetri is situated at the foothills of the Aravalli range in the district of Jhunjhunu in Rajasthan. It is an old town, which came into limelight due to its king Maharaja Ajit Singh's close association with Swami Vivekananda. The copper complex of Khetri is only a few kilometres away from Khetri town. The Khetri mine was developed and started by NMDC and was handed over to HCL after it was incorporated in 1967. Other facilities in Khetri including its concentrator, smelter and refinery were set up later. Khetri consisted of two copper mines, namely the Khetri mine and the Kolihan mine, situated nearby with reserves of about 35 million and 22.5 million tonnes respectively. Next to

Khetri, there was another deposit called Banwas, which had reserves of about 25 million tonnes, but was not developed till then. Both Khetri and Kolihan are underground mines. From underground tunnels, copper ore is extracted and brought over to the surface by using shafts. These two mines together produced about a million tonnes of copper ore pa, which were transported to the concentrator plant for beneficiation.

The beneficiation plant or the concentrator, as it is generally called, had a capacity of 2.4 million tonnes of copper ore pa. Thus, the plant capacity to beneficiate ore was much more than the quantity of ore the two mines together could produce. Even then, the biggest problem that the unit was facing was that it could not beneficiate the ore produced from the mines, not because of the capacity constraint but because of inadequate availability of water. Shortage of water was the biggest problem of the unit. Beneficiation of ore, which is done through a process called froth floatation, requires a large quantity of water. Since adequate water was not available, the performance of the beneficiation plant was always at a sub-optimal level, leading to lower capacity utilisation of smelter and refinery and ultimately loss to the company. The water crisis in Khetri was so well-known and so perennial that it was taken as a given fact that it could not be reversed. After reaching Khetri, on the first day itself, I wanted to understand the situation by holding discussions with all the concerned officers.

Khetri, in any case, is an arid area. It is in the district of Jhunjhunu where about 40 per cent of the total area had desert soil, 36 per cent of land had sand dunes and 8 per cent had red desert soil. Whatever cultivation was taking place in the area was done by extracting groundwater and, as a consequence, in 89 per cent of the total area, the water table was declining steadily over the years because rainfall was never adequate to compensate for the use of groundwater. Incidentally, in India, the average rainfall is about 125 cm pa. In Jhunjhunu, the annual rainfall was less than 50 cm, so it already was a part of the region of scanty rainfall. From 1999 to 2002, the situation was even worse. The area experienced only 62 per cent of its average rainfall of less than 50 cm pa. As a result, there was further depletion of groundwater and further fall in the water table. The scarcity of water, which already existed in the region became more acute, impacting the performance of Khetri copper project even further. Incidentally, there was very little groundwater available in and around Khetri. Water for Khetri used to be drawn from a place which was 40 km away and which had to be accessed through rugged hills and semi-deserts.

I planned to visit that area on the following day itself. I decided to take a four-wheel drive jeep the next morning and reached the place from where water was sourced to directly find out the situation over there. About 20 deep tube wells were installed on a dry river bed from which water was extracted and pumped to Khetri through a 40-km-long pipeline passing through small villages in between. Out of 20 wells, the majority were non-functional because the water table had for quite some time slipped below the tip of the tube and, therefore, the tube wells went dry. Some tube wells were out of order because of other problems that were mainly mechanical. There was another problem too. Whatever quantity of water was available from the running wells, when that was pumped through the pipeline to Khetri, the pipes used to be broken by the villagers for getting water for themselves. Therefore, water crisis became perennial in Khetri and concentrator production was lower by more than 30 per cent which directly affected production of both mines and smelters. I took two engineers with me, spent the whole day in the concerned area and took some simple decisions. Some of the wells could be restarted by expanding the length of the 'be so that the water table which went down further could be touched. We, therefore, decided to reinstall the tube wells with larger tube length and they started functioning immediately. The repair work for the tube wells which were non-functional due to mechanical failures was also undertaken quickly. We decided that, wherever there were villages in between, we would put taps, so that the villagers could take out water from the taps without resorting to breaking the pipeline for extracting water. We knew that it was always possible that taps once opened would never be closed. Therefore, the nozzle size was kept small in order to ensure that, even if water flows out, the quantity of water that will flow would not be beyond a certain limit. In any case, we had decided to request the villagers to close the stop cocks whenever the water was not in use and later we found that the villagers had largely co-operated. All these action plans were executed in just about seven days, and there was considerable increase in availability of water in Khetri. Back in Khetri, we found that it was possible to install two or three more tube wells in Khetri Township itself in order to increase water supply at least to the residents of the township. Those were also installed. It was then observed that there was an opportunity to increase the percentage of water being recycled. The pond storing the recycled water was heavily silted, reducing its capacity. We took a decision to get it de-silted. Simultaneously, the thickener, which was not working, was activated and the percentage of water being recycled increased substantially. These actions produced immediate results. The production in the concentrator plant improved substantially, and by the time I reached Kolkata, production had almost reached full capacity.

Incidentally, the supply of water not only improved the plant performance and output, it also came as a big relief to the residents of the plant township where the residents were suffering in silence due to acute shortage of water. In fact, the households were getting supply of water only once and that too for just about half an hour in the morning. Two deep tube wells that we sunk in the township immediately increased the supply of water to the township without disturbing the plant. If better plant performance gave us more copper, better supply of water gave us better employee satisfaction.

I also noticed in Khetri that the power cost was high because the tariff for supply of power to the township was the same as that for the plant. It was possible to have the connections separated from each other and request the power distribution authority to charge domestic tariff for supply of power to the township. In terms of cost saving, it was not insignificant. I went to Ajmer and discussed the issue with the head of the distribution board. He agreed in principle to do the needful. It took several months, but ultimately it was done. I returned to Kolkata after a couple of weeks with a great deal of satisfaction. I wanted to immediately share the job done with my boss. But he had no time for that. Every time I wanted to meet him, I was told that he was busy and would call me at his convenience, which never happened.

Within the next two to three days, we were called for a meeting in Delhi. It was an inter-ministerial review meeting headed by the Secretary (mines). All the figures presented to him were those for the period ending the previous month, which, inter alia, showed a huge loss of production due to poor performance of the concentrator plant. The Secretary was furious. The chief of operations and my boss both explained that it was inevitable. They said that there was low rainfall in that region and water was scarce. The chief of operations said that it was a very challenging task to run a concentrator plant in a semi-desert area where there was no source of water except rain. He quoted the data on rainfall to explain the situation and expressed the helplessness of the management. The Secretary was annoyed. He wanted to know whether the management had any plan of its own to improve the situation in the given circumstances. I remained silent. But in the end I told him that the situation had already improved. For the last five days, the concentrator was producing almost at the expected level. The water supply had already improved. Certain measures were taken and they were already producing results. I explained the initiatives that had been taken during the previous two weeks. The Secretary appeared to be fully convinced. He only asked how I knew all these things. When I said that I spent two weeks in Khetri, he repeatedly congratulated me. His eyes were sparkling with happiness. I understood its significance later after almost two years.

Meanwhile, the process of disinvestment through strategic sale of 100 per cent government equity had picked up momentum. Out of the two big players in the copper market, Sterlite appeared to be more active in acquiring the company. Sterlite had already secured majority holding in Bharat Aluminium Company Ltd (BALCO). They had also established control over Hindustan Zinc. The company was a natural aspirant to get Hindustan Copper into its fold. The Sterlite team along with their advisors from investment bank came down to Khetri to inspect the assets. They faced vociferous opposition from all the unions. A large number of workmen assembled inside the plant area and started shouting slogans. The team went back. Within a few days, a new date for due diligence was fixed. Khetri had four operating unions, affiliated to the All India Trade Union Congress (AITUC), Indian National Trade Union Congress (INTUC), Bharatiya Mazdoor Sangh (BMS) and Centre of Indian Trade Unions (CITU). While BMS was politically linked to BJP which was in power in Delhi, INTUC was the trade union wing of Indian National Congress, which was the ruling party in Rajasthan where Khetri was located. On the day fixed for due diligence, in the wee hours, some state police personnel came to Khetri and picked up the main leaders of all the unions and took them into custody. Soon after, the district collector along with a huge contingent of the police force took position in different corners of the plant to allow the Sterlite team to complete the due diligence, which lasted for several hours. After the team left, at around nine in the evening, the leaders were released. Both the state and Central Government exhibited their determination to sell off the company and to get rid of the burden of a sick child. Meanwhile, the company's accumulated loss crossed ₹6 billion and the government was too keen to handover the ownership of the company to any willing buyer.

The Sterlite team then came to Kolkata for a discussion with the top management. Encouraged by the helplessness of the government and its determination to sell off the company at any cost, they came out with a new set of demands. They said that they were not getting adequate comfort in Hindustan Copper because the workmen were apparently not in favour of disinvestment, and that they expected workmen, unions and executives of Hindustan Copper to ceremonially welcome them to the company so as to give them adequate confidence to run the company in the way they desired. They added that they had a similar experience in Hindustan Zinc. At least in the context of Hindustan Copper, their expectations appeared to be unrealistic. But it needed persuasion from the management side for them to understand and accept the reality. In such disinvestments, through strategic sale, the process within the company is handled and carried forward by senior executives and top management, but those are the people, who become most vulnerable if the sale materialises. The directors are to quit on the same day of takeover. The executives and managers face uncertainties because their jobs are generally protected for one year only after privatisation and after the initial one year period is over, there is no job security for them that they are accustomed to enjoy in a public sector company.

Workers, on the other hand, continue to have job security as in public sector, because they are covered by the Industrial Dispute Act, 1947, which gives them statutory protection whether they are in public sector or private. When the process of disinvestment acquired momentum, the officers started looking for other jobs. Many left. Those who remained were unhappy and demotivated. The workers were suffering from multiple fears and felt that even if their jobs were protected, they could face unknown dangers. The sick company became sicker in the process. The sale of the company remained the only option. However, while the process of disinvestment started picking up fast, a court verdict on Hindustan Petroleum Corporation Ltd (HPCL) came as a dampener. It stated that if a company or a part of it was nationalised through an Act of Parliament, then the consent of Parliament would be required to privatise the same. This stalled HPCL's disinvestment. It also immediately impacted Hindustan Copper's strategic sale because its Ghatsila unit was erstwhile Indian Copper Corporation, which was nationalised through a bill passed by the Parliament.

4

THE DECLINE

One of the features of a sick public sector company is that it gets too many advisors, mostly from the administrative ministry. They advise the management on almost every matter, small or big. It starts with the anxiety of the officials in the ministry regarding the health of the company and the general belief that the management of a sick company is either less efficient or less equipped to handle the issues the company is confronted with. Slowly, advice becomes instructions and instructions become directives. When a sick company approaches the administrative ministry for funds or other support, the officers who process the proposals feel that they need to be fully convinced about the justifications and start asking questions and offering suggestions. They feel that as the owner's representatives, they have both the right and the duty to ensure that the PSU is properly run, though managing a company is neither within their field of specialisation nor their job. The top officials of sick companies, however, try to keep the officers of every level happy in their anxiety to make their proposals moving. Soon the interventions go beyond the specific proposals and cover almost all aspects of company's functioning. Sometimes, the company management takes advantage of the situation.

They coolly pass on the job of managing the company to the government officials who are asked to give their views and directions on all matters, including those which are inconvenient. Very few government officers can understand the strategy and fall into a trap, which they discover only when they receive proposals with 'as advised' notes for approval. This is the general trend in sick PSUs, and Hindustan Copper was no exception.

Once a fire broke out in a sub-station near Khetri. The place was about 40 km away from the Khetri plant, but that was the main power intake point for the unit. The entire plant sank into darkness. Production stopped. I was advised to rush to Khetri to handle the situation. I went to the spot and found out that it was evidently an accident and there was no apparent reason to ascribe it to human negligence. We were trying to restore the power connection as early as possible by repairing the sub-station and also by drawing power from other sources. The latter effort gave immediate results and the power supply resumed to the township and office building. However, it was not possible to start the smelter. After two days, the CMD of the company came to the site and told me to prepare a report to be sent to the ministry, which I did. He was not happy with the report. He wanted to highlight two aspects. First, the loss of production, which had taken place, and further loss, which would happen till the power supply was restored. Second, he indicated that there would be a thorough inspection by a team of technical experts to identify the officers at fault and that appropriate action would be taken accordingly. I argued with him against the incorporation of these two points in the report. He said that he did not believe in giving incomplete reports to the ministry, which ultimately was the owner of the company. The report went to the Honorable Minister. The minister, a young man with a legal background, was furious on receiving the report. He reprimanded the Chairman and said that the company was already making losses and at this juncture, if the officers were responsible for the fire, which caused loss of production, most severe action against the concerned officers should be taken. He also said that it would be construed to be a failure on the part of the Chairman if he failed to take such actions within the shortest possible time. The Chairman, meanwhile, identified a technical expert who was flown in from Hyderabad to Delhi and then sent from Delhi to Khetri to carry out the technical enquiry. Within a couple of days thereafter, the Honorable Minister came to Kolkata and called the Chairman to his hotel room. I accompanied the Chairman. The minister was still in a bad mood. He said that the conduct of the officers who were responsible for the fire was terrible and that they should be given exemplary punishment. The Chairman explained that he would get the enquiry report within the next few days and take immediate action. The minister was still angry and desired to know the extent of loss the company might have incurred. At that stage, I intervened. I told the minister that we were continuing the production by importing copper concentrate and whatever be the compulsion for carrying out the production, we were incurring loss on every tonne of copper cathode we were producing. Due to the smelters being shut down, the company lost production, but not profit. In fact, the amount of loss had come down. The minister was relieved. He smiled for the first time during the discussion and said that, nevertheless, if the officers were responsible, they should be punished. The enquiry report came after two-three days and the technical expert clearly indicated that nobody could be held directly responsible for the fire which was purely accidental. However, two or three officers who were already placed under suspension for no reason whatsoever continued to remain under suspension' until clearance from the officers in the administrative ministry was obtained. I still fail to fathom why the Chairman provoked the Honorable Minister who reacted on a report which could have annoyed anybody. In my understanding, it was only because the Chief Executive Officer (CEO) wanted to distance himself from other managers of the company and prove himself to be a tough manager. He possibly felt that he could adopt a coercive approach with his officers under the garb that it was the desire of the ministry or minister, using the ministry as a shield as was the common practice in sick PSUs.

Often, the ministry is fed distorted logic, biased information and wrong analysis. Based on the data furnished by the company and the discussion with the company's top management, the ministry officials came to the conclusion that Hindustan Copper was losing more on mining operations and, hence, the mines were to be closed and smelters expanded. They felt that if the smelters were modernised and expanded, it would reduce the loss of Hindustan Copper. The problems of mines on the other hand were incurable because the global price of copper was low and the cost of operations was high. It was decided that more loss-making mines should be identified and closed. All the mines of Mosabani area in Ghatsila were included in this list and they were shut down one by one. Next came the turn of two other mines in Ghatsila, namely the Surda and Rakha mines. They were also decided to be closed. A proposal was simultaneously moved to expand the capacity of Khetri smelter by about 16,500 tonnes and Ghatsila smelter by about 7,500 tonnes pa. That would expand the smelting capacity of Hindustan Copper to about 71,500 tonnes pa, from 47,500 tonnes as was the capacity. After the closure of mines, production of copper ore and concentrates came down to about 32,000 tonnes of copper in concentrates (metal in concentrates/MIC) and the idea was to import more copper concentrates to bridge the gap. The government decided to fund the capital investments required for expansion of smelter capacity, and proposals from the company were mooted accordingly.

When deliberation on these issues was going on, suddenly, an advice came from the topmost level of the ministry directing that Khetri mine in Rajasthan, which was making loss, should also be closed. It was surprising because Khetri was a marginal mine, and if Khetri was closed, a part of the overheads would be loaded on the other mine called Kolihan in the same area, and with additional overheads being loaded, Kolihan would also become uneconomical. We decided to argue against the decision, but then we got a bigger shock. We came to know that it was more a political decision than an operational requirement. It was politically thought to be necessary, though I do not understand even now why it was to be so, to close a mine in Rajasthan in order to offset closure of a number of mines in Jharkhand just because the two states were ruled by different political parties. It was true that Khetri was making loss, but much of its losses could be attributed to the exceptionally low price of copper at that time. On the other hand, construction of any underground mine like Khetri would take many years and would cost millions of rupees. If it was a choice between abandoning such a mine and operating it by bearing marginal losses expecting the price to be corrected soon, the latter would be the obvious choice. (In fact, within three years, the copper price became fourfold and Khetri became an asset for the company). However, no logic was heard, no opposition was entertained and Khetri was decided to be put under closure. The matter was directed to be put up to the board. Some of the independent directors expressed reservations and were persuaded to agree. The closure notice was issued and the employees were given an opportunity to apply for voluntary retirement in order to avoid retrenchment under the law. Out of about 1,100 workmen, the majority opted for voluntary retirement. Unions went to the court and obtained a stay order restraining the management to retrench the remaining work force. As a consequence, about

400 workmen remained in employment, but as the mine was closed, they had no job to perform. They started getting idle wages. The company hired good lawyers, but the judges appeared to be convinced from the records put up by the unions that the company did not have any conclusive reason behind the decision to close the mine. The company landed up with more losses by closing the mine than what it incurred when the mine was in operation.

After Khetri, the focus came back to the eastern sector. In Ghatslia, all the mines except two mines, namely the Surda and Rakha mines, were already closed. Closure of those mines was logical because the mines had already completed their productive life. Most of these mines were more than 1 km deep and the operation was fully uneconomic. Even otherwise, the reserves that these mines were left with were meagre and by closing those mines, in effect, no economic resource of the nation was being abandoned. The situation in Surda and Rakha was better, but unlike Khetri, which was making marginal loss, the loss generated by these two mines was much higher. The annual production capacity of these mines was also low. While Surda could produce just about 250,000 tonnes of ore in a year, Rakha's capacity was even lower. It could produce just about 100,000 tonnes of copper ore pa. The metal content of the ore produced from these two mines was about 250 tonnes and 100 tonnes pa respectively, and the prospect of these mines being made viable was remote. Moreover, Surda did not have a concentrator plant of its own and, therefore, the ore extracted from the mine had to be transported to the concentrator plant at a distance of about 8 km, which increased the cost of operations further. We had a number of meetings with the unions. We had lengthy discussions with the executives and, after exploring all the possibilities, we found that the Rakha mine should be immediately closed.

Meanwhile, the government came out with another idea of shutting down the company's headquarter in Kolkata and shifting it to Delhi. The plan was to sell off the corporate office in Kolkata, which was to fetch about ₹250 million. This decision had a huge impact because almost all the employees were to move out of the company on voluntary retirement, because with unrevised salary and delayed payment, it was almost impossible for most of the employees to relocate to Delhi. Despair came down in the corporate office resulting in agitations by the employees. Hindustan Copper's headquarter in Kolkata had the best possible work culture in terms of discipline, performance and attitudes, which were different from those in many other offices in Kolkata. The employees started agitating during lunch breaks. The government officials on the board blamed the management for that and suggested that stern action should be taken against them. When asked about it, I decided to be blunt. I told the Board that if there was a secret ballot, 100 per cent employees except the Chairman and two other directors would vote against shifting of headquarter to Delhi. These three will not vote because being on the board they had no other option but to go by the government's decision. Given an option, they would also vote against shifting. I said that at a critical time, when the company was passing through the worst phase of crisis, shifting of corporate office and dismantling the corporate team would be unwise, and that too for such a small amount of money which was not worth the trouble the company would face. For the first time, the government representatives on the board remained silent. My other colleagues on the board nodded in consent. The issue was not pursued further during the board meeting, but it remained alive and pressure came from the topmost level to implement the decision which, however, was resisted. This was a phase when I came across the highest degree of insensitivity from those who apparently were trying to help the company from the government. Some of the comments made during the internal discussions were just unthinkable. We were advised to close down all hospitals of the company to save cost. When it was argued that in the remote areas where the mines were situated, closure of hospitals would create more problems than solving them and by the time the critical patients were shifted to the nearest town, some of them would die, the response was an immediate, 'Let them die'. It was justified by the argument that Indian workers had low productivity. In terms of productivity, 20 Indian workers were equivalent to 1 American worker. So the country would not lose much if Indian workers died. We were fortunate that such insensitivities were limited to a few and not widespread.

As time passed, the situation went from bad to worse. The company's cash flow in particular was entering a dead end. The banks demanded a government guarantee for the entire amount of cash credit limit. Though the government guarantee was not free of cost, yet the government was reluctant to offer it. The cash flow position entered a critical stage and the salaries were delayed further. The company in the meantime defaulted on the repayment of money it had raised through bonds at very high rates of interest ranging from 14.5 per cent to 15.75 per cent. The company continued to pay interest, putting further load on its cash position. It was apparent that the company could not continue like that. The directors retired, but were not replaced. There was no initiative on the part of the government to fill up the positions of functional directors on the board which were vacant due to superannuation or resignations. The post of Chairman was going to fall vacant at the end of November 2003, but no initiative was being taken to fill the vacancy. I was left to be the only functional director in the company. There was a clear indication that the owners of the company, having failed to sell the company, were preparing to close it down. Then came 29 November 2003, about which I have already narrated at the beginning of the book.

5

ACTION BEGINS

When I took charge of the company as its officiating CEO on the afternoon of 29 November 2003, I knew the extent of problems the company was facing and I knew the gravity of challenges I was to handle. But when I decided to take the bull by the horn, I got to know more about the seriousness of the issues that the company had been undergoing. I already knew that Hindustan Copper was to be put under care and maintenance with all the operations suspended with effect from (w.e.f.) sometime in December 2003. That was the reason why no initiative was taken to appoint a new Chairman and the existing Chairman was proposed to be given three months' extension, which, as I stated earlier, was turned down by the appointment committee of the cabinet. The situation, which required the government to take a decision to allow the company to suspend operations, was that in the month of December, the payments that were necessary to meet the inescapable liabilities stood at ₹750 million, while gross collections and available cash taken together would not exceed ₹200 million. LCs opened by the banks in favour of Hindustan Copper for the import of copper concentrate had already been extended beyond 500 days.

It was no longer possible to extend those LCs and defer payment. The company's bank accounts were on the verge of getting frozen. The consortium of banks, which had fixed the cash credit limit as ₹1.65 billion, declined to increase the limit and insisted on a government guarantee even for the existing facilities. Default on payment of the power bill had come to such a stage that it was almost impossible to avoid disconnection. At ground level, production was already down. The Khetri mines had already been closed. Out of the 1,100 people working in the mines before the closure, about 700 people opted for voluntary retirement. The unions went to court challenging the decision of the company to close down the mines and secured a stay order from the court. Consequently, about 400 people who did not opt for voluntary retirement remained on the rolls of the company and had to be paid salary though there was no production from the mines. There was no import of concentrate during the previous month and mining production was low. Consequently, production of copper cathodes and copper rods, which were marketable products, was quite low. The marketing offices in various regions had already been closed after almost all the marketing executives were released through the voluntary retirement scheme. The market share of Hindustan Copper had already been very low. Sales revenue was also one of the lowest. In November 2003, the company had incurred a loss of about ₹100 million. The company was left with about 6,000 employees, with low productivity and low per capita value addition. In fact, the value addition per rupee of wages was just about one and a half rupees. The morale was low and obviously the atmosphere was full of uncertainty, apprehension and anxiety.

After the CMD superannuated, the company had only one functional director, that is, myself who was the CMD, director–personnel, director–operations and director–finance, all rolled into one. In a crisis situation, this gave me an unusual feeling—a feeling of liberation. I felt that if I had to act fast and in unconventional ways, there could not have been a situation better than this. What I thought could easily be converted into decisions, and what I decided could easily be put into action. There was hardly any need for processing time. This might not have been an ideal situation under normal circumstances, but in a situation of crisis, this is what was needed. The company was sinking every day and the need of the hour was to think, plan and act within the shortest possible time to keep it afloat and draw it to a safer zone at the earliest. If we could do so, we would become more participative, which would help us consolidate and improve.

My first task was to draw up a quick plan, a plan that would keep the company afloat. Certain actions were taken with utmost urgency. They were most important for the survival of the company. The company was sinking fast and approaching a stage of hibernation from which nobody knew whether it would come back to life or not. The immediate requirement of the company was cash and, therefore, the first and immediate task was to collect cash as early as possible and to release it as late as possible. Management of cash in the company was the first task. It was not, however, easy. The company did not carry sundry debtors because of the system of sale that the copper industry in general followed. All sales were made against advance payments which were to be adjusted towards the physical sale of materials during the entire month. Generally, the advance payments collected from the customers were more than the value of the material delivered to them, because the monthly copper price was determined on the basis of the average of daily price fixed as per the price announcements made by the LME. The sellers in the market also, therefore, always kept a small buffer in accepting the advances, lest the average price for the month exceeded the estimated price based on the current trend. So, the only way to generate cash was to sell more and to get more advances. It was a huge challenge because sales had already dipped. Most of the customers had already deserted the company and even the stock position was not comfortable. Avoiding payment was equally difficult, if not more. The most important was to defer payment to the bank for LCs drawn about one and half years back for import of copper concentrates. The company was threatened with the freezing of all its bank accounts if payment was not made. But then there was no money to make payments even if the cash collection improved. However, equally important was the release of some payment, especially on electricity and some other pressing dues, as otherwise the entire operations of the company would come to a grinding halt. Therefore, actions were needed to be taken to address the issue of cash management on top priority.

The second task was related to the first. That was to generate confidence among officers and workers at every unit and at every level that the company was going to survive. A message was to reach everybody that the company was not heading for collapse, all hope was not lost and there was reason to believe that it was clearly possible that the company could come back from the brink of disaster. It was also necessary to communicate that all that was needed was the dedicated effort of people at every level. Whatever little bit of the regional sales offices that did exist needed to be strengthened and activated to grab orders and provide confidence to the customers that the company was not going to be sunk and that they could rely on the company as they used to do in the past. It was necessary to give confidence to the employees working in the shop floors that, despite all odds, production was to be improved and the systems were to be strengthened. The message was to go around that the only way to survive was to produce more and sell more, and that there was no other alternative.

The third task was to reopen the closed mine in Khetri, where most of the workers were separated through voluntary retirement scheme (VRS). The remaining workers were getting idle wages due to the intervention of the court, which had granted a stay on the closure. Due to the closure of the mine, availability of minerals in Khetri had drastically reduced. Only Kolihan mine was being operated in the same region, but the loss of Khetri mine not only led to under-utilisation of the capacity of the concentrator plant which was set up to treat the ore received from both Khetri and Kolihan mines, it also resulted in under-utilisation of the smelter. The next plan was to stop the import of copper concentrate despite an advertisement issued sometime back to that effect. It was clear that the import of copper concentrate was ultimately detrimental to the interest of the company.

The fifth plan, which was a corollary of the other plans, was to run the smelters on full capacity by using only the internal production of copper concentrates of the company. In other words, the plan was to operate only one smelter at a time, so that it could give full production and not to operate two smelters with lesser capacity utilisation. For this purpose, Khetri smelter, with a production capacity of 31,000 tonnes pa was the immediate choice. The small smelter at Ghatsila was also kept in working condition so that during any shut down in Khetri, Ghatsila smelter could be used in full or more than 100 per cent capacity.

The sixth plan was to sell scrap to the extent possible. It was immediately identified that some metal scrap was available in Khetri, after a decision was taken to close down the fertiliser plant sometime back. An arrangement was made to sell it immediately and to collect whatever cash it could generate.

Finally, it was also necessary to start working on a long-term plan to strengthen the fundamentals of the company and to ensure that its revival is sustained. While all these plans were drawn by the end of 29 November 2003 itself, a very important task still remained; we needed to organise a warm farewell to the outgoing Chairman of the company.

I had a lot of differences with him and many a times, some of the differences crossed the professional level and intruded into personal relations. To a large extent, it was my failure. On introspection, I could clearly see the many qualities the gentleman had.

He was handling an extremely unenviable task of managing the company at a stage when it was passing through the most difficult stage of its existence. He was a man of wit. He was extremely intelligent and articulate. Only his way of looking at things were different from those of mine. Some of my initiatives might have hurt him personally. I never felt guilty for that, but I obviously felt guilty due to the fact that I could not maintain the warm personal rapport with him which I always enjoyed with all my superiors whom I had worked with in the past. I felt that no initiative for reviving the company could succeed if his contributions were not recognised and he was not given a very warm farewell from the company, which he genuinely deserved. I called him up late in the evening, shared with him the developments and requested him to come down to attend the farewell the next day. He agreed. I along with all the officers of the company who were present in Kolkata gathered at the guesthouse, received him with flowers and spent a lot of valuable time with him, learning from him what he had done for the company and how he would have liked to handle the situation. We shook hands and embraced each other. Almost at midnight, when the farewell ceremony was over, we all felt that morning was not far away and we must jump into action without delay. No discussions on the past was required. Blaming anyone was irrelevant. No repentance for any lost opportunity was meaningful. What was important was only to act, and act with determination. We knew that we were in a dark tunnel at the end of which we were sure to see the light. While walking in that direction, I in particular felt much light-hearted having buried all the bitterness of the past. Free from burdens, I decided to walk quickly and determinedly in the direction of light. Many years later, I met the outgoing Chairman in some other town of India. I was extremely happy to have been received by him with a lot of personal warmth and to have enjoyed a fabulous dinner together.

As stated, the first problem that was to be addressed was the critical cash position and obligation to pay for the LCs issued by the State Bank of Bikaner and Jaipur. We had no money nor did we have any hope of getting any assistance from any quarter.

We only discovered that sometime back the government had sent a communication to the company committing to the release of some funds at the end of the financial year, that is, sometime in March 2004. Armed with this letter, I decided to meet the managing director of the bank and do everything possible to explain the position to him. I landed in Jaipur within the next two days to meet him. The managing director was a nice person, a patient listener and sympathetic. He expressed his total helplessness. He said that the board of his bank was annoyed at him for allowing the LCs to be extended on so many occasions covering a long period and he had hardly any option left. I took advantage of his positive attitude and explained to him that if his bank took the actions as planned against Hindustan Copper, both the organisations will land up in a lose-lose situation. The bank will not realise the money and Hindustan Copper will go into hibernation. I assured him that, latest by March 2004, the company would be in a position to clear the dues and the bank could be sure of it. However, if the bank did not show patience, the company would be closed and the bank would not get its money back. I took out the letter of the government and said that what

I was telling him was on the strength of that letter. I assured him that if he co-operated, the company would turnaround and would be in a position to handle the liability on its own strength. Nevertheless, the government funds were in any case available latest by March and there was no risk for the bank. The logic sounded convincing. The bank chief was apparently impressed by my sincerity and conviction. When he ordered a second round of tea, I knew that I was not fighting a losing battle. He agreed that he would be advocating our case as strongly as possible before the board to get its clearance. Ultimately, he convinced his board and the LCs were extended. The company was able to avoid two major problems, that is, (a) its bank accounts were not frozen, and (b) a large part of the liability which was due to be cleared in the month of December could be deferred, giving a much-needed breathing time to the company. The gap between ₹700 million liability and ₹200 million revenue was bridged to a substantial extent, making my task easier. Never in future did I forgot the help and support I received from this person.

....

> I not only ensured that by March 2004 all the dues were cleared, even later, when the company started doing much better, I kept the gentleman informed about the progress made by the company and reminded him of the invaluable support which he had given to the company, but for which the company could not have survived and turned around.

....

I then decided to reopen the issue of closure of the Khetri mine, where 400 workers were being paid idle wages on court orders. Further, due to loss of production arising out of mine closure, the smelter was operating at a lower capacity, resulting in a higher cost of production and lower sales. The neighbouring Kolihan mine started showing higher cost of production due to shift of overheads from the Khetri mine to Kolihan. I considered this to be the most important issue to be taken up. I went to the ministry, explained the situation to the additional Secretary and argued that we should run the mine even with 400 people and generate some production from the mines until the court case was resolved, because there was no point paying salary to the people without giving work to them. The proposal was turned down on the ground that it would not only mean flouting of a government decision but would also weaken the company's position in the court. It was about 7.00 PM when I came out of Shastri Bhawan. I immediately decided to take the next flight to Jaipur and meet the lawyer representing us in the court. I landed in Jaipur at about 10.30 PM and drove straight to the lawyer's house. We sat together for the next one and a half hours. He agreed with me that there was no risk of the company's case being weakened if the mine was reopened. I persuaded him to give his opinion in writing. He agreed. I came out of his house well past midnight and the next morning at 9.30 AM was back to Delhi to resubmit my proposal along with the legal opinion. I was told after three hours that it was not possible to approve the proposal because that would in any case violate the government decision, but the Chairman of the company was free to take his own decision. The decision was obvious. I called the workers' representative immediately to Delhi and told them that the mine could be reopened, but they would have to operate it with 400 odd people. Only in some critical areas, a few workers would be reengaged on contract, but the mine should give normal production. The workers and union leaders were excited and more than ready to co-operate. They never expected this. They said that 400+ workers who were still on rolls would accept multitasking and ensuring that the mine produced at full capacity. Within the next couple of weeks, the mine started production. There was celebration all around. The employees of Khetri mining complex as well as the people of the neighbouring areas in the district of Jhunjhunu celebrated the decision. In thousands of houses, people lit lamps and prayed for the prosperity of the company. I was told that they even prayed for me, seeking God's blessings on me, which was most important for me in my battle against so many odds. In January 2004, the Khetri mine started giving profits. A few months later, when the issue came up for discussion in a review meeting, the issue of implementation of the earlier decision for closure of the mine came up once again. I was already armed with the strongest possible argument that the mine was giving profit. I stated that as the Chairman of the company, I was responsible for its profitability and no government decision was directly applicable to me unless it came through a Presidential directive. The debate and discussions continued for quite some time, but no conclusions could be drawn. I could understand the position of top government officials. They could not take a view, which was opposed to a decision from the top. Simultaneously, they knew that closure of the mine when it was making profit would be unreasonable. Only the representative of the Planning Commission clearly supported my view. By that time, the Ministry of Coal and Mines had a new minister who ventilated her opinion in the matter, though indirectly, after a few weeks, and as a result, the issue never came up for discussion ever again. The Khetri mine continued to be alive, and it remains operational and profitable even today.

Two major battles had been won during the first week of December itself. But then worse battles were still to come. During the previous few months, the company had lost most of its customers. They had all been snatched by two major private players operating in Indian copper industry. Most of these customers were loyal to Hindustan Copper for many years, but all communication with them had snapped after the regional offices were closed down. The western regional market which consumed most of the products of the company was the worst hit. In the second week of December,

I camped in Mumbai for two to three days and met the customers one by one. Copper prices being regulated by the daily rates announced by the LME, there was hardly any scope to give any discount to customers. In fact, the prices of various copper products were mostly standardised. Since the prices changed daily, the transactions were done on average monthly prices and until the end of the month, all payments received were provisional and were adjusted with the final price derived on the basis of the average of daily rates for the whole month. Due to this system, customers were tied to the supplier more firmly. There was always some excess amount paid by the customers that was adjusted with the next month's purchase. It was, therefore, difficult to get the old customers back. The task was more difficult for Hindustan Copper, which was a smaller player in the market and was carrying the stigma of sickness. I decided to approach the old customers personally to win them over. It was a difficult task but some of the medium-sized customers agreed to lift material from us once again. But some big customers expressed their reluctance because they had no commercial reason to shift their loyalties from the existing suppliers. I understood their position, but continued the discussion. One of the largest customers of copper rods wanted to meet me in Willingdon Club, Mumbai. As an ever-obliging salesman, I went to the club to meet him. He had a very interesting proposal to offer. He suggested that he would buy material from Hindustan Copper, if we agreed to sell at the rate prevailing on the date of delivery and not on the average monthly rate. This was not generally done because if the rates went up during the subsequent period of the month, the average rate for the month would exceed the daily rate and the company would suffer loss. It was just impossible to take this risk, especially in a public sector company. However, it was an opportunity to improve sales and increase the cash collection during a month, which was very critical for the company. I decided to agree, but did not disclose my plan to him. I just went out and quickly telephoned the other government-nominated non-executive members of the board, namely the joint Secretary and the additional Secretary of the ministry and told them that I was agreeing to this proposal and I needed their moral support. Both of them, from different locations, told me that they were convinced that whatever I did would be in the best interest of the company and they would ratify the decision in the next board meeting. I went back to the table. The cup of tea, which was laid on the table, was already cold. The customer ordered for a fresh cup and to his utter surprise, I announced my decision agreeing to his proposal. This time he was taken aback. He never could expect that a public sector company would agree to his proposal, which was best suited to him at that stage. I returned to Kolkata. From the next day onwards, I developed a habit of checking the LME price of copper first thing in the morning. During the next couple of days, the price increased, I lost a few heartbeats, but then the price fell and fell steeply till the end of the month. The company not only earned a handsome revenue from this transaction but got an additional ₹5 million as price differential. The following week, the board meeting was fixed. One of the items of the agenda was to ratify this particular transaction. During the board meeting, the additional Secretary told me that she could not sleep well during the last two weeks and she was also checking the copper prices as I did. I do not know now as to what would have happened had the copper prices gone up and the company had lost money. My subsequent experiences, however, have convinced me that nothing miserable would have happened.

> If the intentions are clear, it is not difficult to explain and come out clean. There can be some difficulties for some time, but the satisfaction of taking unconventional decisions and tasting success is worth risking those challenges.

6

NEW STRATEGY

After the initial hectic days, I was convinced that the company would not be shut down. There was no question of the company being put under care and maintenance. The projected cash flow of December was encouraging. The State Bank of Bikaner and Jaipur had agreed to extend the LCs. The sales revenue was expected to be more than what was initially projected, thanks to the adventurous decision taken at the Willingdon Club, Mumbai. Some production from the Khetri mines had just started coming in. The concentrator plant was performing well. I could clearly visualise that with the aid of effort and luck, we were just going to avoid the most critical crisis, which seemed unavoidable only a couple of weeks back.

I always felt that Hindustan Copper was following a wrong business strategy. It was a mining company and mines were its assets. The greatest value of these assets had been that no other copper manufacturing company in India possessed such assets. The company's smelters were only to do value additions to the ore mined by the company and the copper concentrate it produced. These smelters were not capable of competing with the huge shore-based smelters of the two large private sector manufacturers who used to import copper concentrate and do smelting and refining in their large modern plants to produce saleable copper. The fact that Hindustan Copper was losing more on mining operations than on smelting and refining should not have meant that smelters were more important assets of the company than mines. It was only with mines that Hindustan Copper could maintain its competitive edge over its competitors, who had to import copper and did not have mines from which to obtain copper concentrate. The concept largely shared by the government officials that the smelting capacity of the company should be expanded through capital investment was not right because the existing copper smelting capacity was more than what the mines of Hindustan Copper could produce. It appeared to me that the import of copper concentrates was commercially an unviable initiative, because with imported concentrates, it was impossible to compete with private sector smelters. Already the cost of production in our smelters was high. It also involved a huge cost on logistics. Every tonne of copper concentrate was to travel from Kandla to Khetri for smelting, and then the cathodes would travel from Khetri to Taloja in Maharashtra, and from there the copper rods would be transported to the customers in different regions. This was not a viable model of production. On actual processing cost, the small-sized smelters in Khetri and Ghatsila, which were constructed long time back, could never be competitive with the cost of production of the modern shore-based smelters of private sector players. The biggest advantage of the smelters of Hindustan Copper had been that they did not have to import concentrates as was done by the private copper manufacturers. They could get concentrates in the same area from the mines and concentrators and, therefore, with a little bit of effort, they could get much cheaper concentrate and convert them into copper cathodes at a reasonable cost despite the higher processing cost of those smelters. The position became so clear to me that I immediately told the administrative ministry to not even think about making any further capital investment on expansion of smelter capacity. The ministry officials were surprised because that was the proposal which had been pushed vigorously during the preceding two years. It was, however, a relief for them because they themselves had found it difficult to push the proposal further in view of the poor performance of the company. I decided that instead of two, one smelter at Khetri would operate so that it could have full capacity utilisation from the production generated from the mines. The other smelter at Ghatsila was to be kept as a standby and operated only when the larger smelter in Khetri had to be shut down for periodical overhaul. I anticipated some resistance from unions, but was happy to observe that they understood the logic equally well. The model of operation was clear: operate the mines at full capacity, operate smelters capacity to the extent raw materials from the mines were available, stop import completely and market the product directly in the regions where customers were getting confidence back on Hindustan Copper.

> The essence of this model was that Hindustan Copper was to be managed predominantly as a mining company where smelting and refining operations were only for post-mining value-addition activities.

With this model in place, the next task was to increase the mining production by removing the bottlenecks wherever they arose in the past. Hindustan Copper was the owner of all the copper mines in the country and the Malanjkhand mine was the largest copper mine in India. It was an open cast mine and was contributing about two-thirds of the total production of the company. By itself, the Malanjkhand copper complex was a profitable unit. But then, due to a huge cash crunch, the productivity of the mine was declining because the job of overburden removal and mine development was deferred. We decided to take up the job within the next few months and put whatever resources we could generate from our operations into that. Production from the Khetri and Kolihan mines was also planned to be increased. It was also planned to start extracting minerals from a copper reserve called Banwas, which was adjacent to the Khetri mine through the same shaft used by Khetri. But though these plans were finalised immediately, these were to take time for implementation. In fact, in subsequent months, much of the attention of the management was given to these areas and the company got long-term benefits out of these initiatives.

However, the immediate problem that arose from the implementation of this new strategy was that the availability of saleable copper became less than the smelting capacity. As a result, the market share of Hindustan Copper declined, though with lower production, the financial position of the company became much better and every tonne of copper produced started giving comfortable earnings before interest, taxes, depreciation and amortisation (EBITDA).

By the third week of December, I could feel that a miracle was underway. Initially, I was expecting that we would be able to avoid suspension of operations, and it was now slowly becoming apparent that something unbelievable was about to happen. The possibility of posting a net profit after 90 months of continuous loss started looking like a reality. Even in November 2003, the loss was about ₹100 million. In December, the picture was going to be different. On 21 December, the then Secretary of Mines, came to Kolkata. I received him at the airport. The state government, as per the protocol, had sent a Contessa car to the airport. The car was old and the two of us were travelling in that car from the airport to the city. After a few minutes, the gentleman very apologetically enquired when exactly I was planning to suspend operations and if it was really possible to release payment of at least one months' salary to the employees. He said that he was sad, but he did not have a way out because any financial support could be available only in March 2004 and not before. I was internally amused, but remained silent for a few seconds and then broke my silence. I stated that I was going to tell him something which was too good to believe but too exciting to conceal. I told him that the company was not only going to run, it was going to post a net profit after 90 months of continuous loss. I looked at his face. It was a mixture of surprise, disbelief and jubilation all in one. I wanted to reconfirm it by reiterating what I told him before. His face brightened. He was a man of tall structure and heavy physique. He jumped in excitement. I feared for a moment whether the old Contessa car and its springs would withstand the physical expression of his joy. I looked at his happiness and the extent of relief that his face revealed and felt that all my efforts had just received its first reward. From that day onwards, till date, the gentleman has remained a great friend of mine and a great admirer. The public expression of his admiration, whenever we happen to meet, makes me embarrassed even now.

Another interesting incident occurred in December. On 23 December 2003, I went to Delhi and met the Honorable Minister of Mines. I took an appointment and went to his house. The minister, I knew, was a little upset after his recommendation for granting three month's extension to the former CMD of Hindustan Copper was turned down on the very next day by the appointment committee of the cabinet. I entered his bungalow and met him in his drawing room. He was alone and boisterously told me to come and be seated. The conversation which began went something like this.

> Hon'ble Minister (H.M.): So, at long last you have got time to meet me.
> Me: Sorry Sir, but I was only doing the work that you have assigned to me. I am just trying to do whatever I could do for your company Sir.

H.M.: I see. I know you have taken over, but then there are a lot of complaints against you.
Me: Sir, is there any complaint now or you are talking of any complaint that you may have heard in the past?
H.M.: What do you mean?
Me: Sir, there could be some people who could have complained against me in the past, but now there is none who will complain.
H.M.: Hmm, you see, my officers wanted to put somebody else in your place, but I decided to put you in charge.
Me: Why Sir? If somebody else would have been brought in, that would possibly have been better, because then two of us could have managed the company instead of my doing it alone.
H.M.: No, I wanted you. Work hard and try to do good for the company. I want you to perform well. I think you have issued an advertisement for import of concentrates. When are you going to get it?
Me: No Sir, we have decided not to import concentrates because that is not economical.
H.M.: I do not understand. How then will you maintain your production? Why did you then issue the advertisement to import? Is it not wastage of money? Yours is a sick company. Are you going to make it more sick?
Me: I can assure you Sir, that you will soon see what under your leadership we are going to achieve. For the last 7 years the company has survived only on Government support, but Sir, if my assessment is correct, we will post a net profit this month after 90 months of continuous loss. What had been your source of shame would soon be your point of pride. There will be a turnaround of the company and it will be on profit.
H.M.: Sounds unbelievable, but it is just a matter of 7 days more. I will wait.

On 31 December, I sent a flash report to the ministry, indicating the performance of the company. Next day, I got a message that all the officers working in the Ministry of Mines were advised to attend a meeting on 6 January to be chaired by the Honorable Minister himself in which I was to make a presentation regarding the initiatives that made such an unbelievable turnaround possible. I realised that the Honorable Minister meant what he said. He was really waiting to see what was happening in Hindustan Copper and took pride in what happened. However, on 5 January itself, there was a reshuffle in the ministry, and a new minister took over. The presentation never took place. But the changes happening in Hindustan Copper did not skip the attention of the new minister. I got a letter of appreciation from her within the next few days. The journey towards turnaround had just begun. There was a long way to go before the gains could be consolidated and the company could firmly be placed on the path of revival. The struggle continued.

7

SETTLING IN

The first month after I took over was hectic, though rewarding. As I stepped into the new year, I felt that it was time when one should look within and consolidate the position while continuing to pursue the other initiatives which were necessary. The workers and officers were slowly gaining back their morale, but then it was important that the company's position be properly communicated to them.

> I always believed in the fundamental definition of an organisation, which is nothing but the sum total of people who work towards attaining a common goal. This is the beginners' lesson in management, though often forgotten.

An organisation exists for people, is run by people and reflects the people's motivation, expectation, agony, zeal, frustrations and all other sentiments. I have always felt that people, if properly motivated, can achieve miracles, of which the ultimate beneficiary is the organisation. I never felt that an organisation should be run only by rules. To me, rules are codified common sense and laws are codified experience. I have seen that in many organisations, the rules are treated to be so sacrosanct and so rigid that when rules come in conflict with people's aspirations, the rules prevail. The best of employees quit the organisation because the rules do not permit certain minimum benefits expected by them. Many a times, people suffer because whatever they need may not be permitted by the rules.

> To me, it always appeared that an organisation which is not run by its managers but by its rules tends to become a stale, faceless, and mindless structure and can never give the desired results.

I have also seen that communications within the organisations are often so ridiculous that even a worker who is known to have no knowledge of English language receive communications from the company in legalese English which might have been drafted with the help of lawyers. An organisation belongs to a worker too, and if any communication to him/her is necessary, the purpose of it is defeated if the communication is not understood. Many a times, the human resource department in public sector companies in particular, instead of helping the company by evolving a realistic method of working, creates a false structure, which helps nobody. In my life, wherever I have worked, I always felt that if there was a conflict between the rules of the company and the genuine needs of the employees, the rules should be amended, modified or even broken. What really matters is to ensure that the employees' needs are properly addressed, because it is the employees and not the rules who make an organisation dynamic and responsive to different situations. Let me mention two incidents associated with this.

It was January 2004, and I was busy in a review meeting. A union leader came to meet me and very apologetically requested for some help for a worker of the lowest rung whose son was admitted to Christian Medical College, Vellore. The boy was suffering from leukaemia and needed bone marrow transplantation. The cost involved was about ₹1.3 million. I told a colleague to immediately check the rules. The rules were silent, and if we tried to interpret, there were more arguments in favour of denying the request than accepting it. I thought for a while. It was impossible for the concerned worker to really arrange funds for the surgery. In his zeal, he had somehow taken his son to Vellore, but now found it impossible to bear further expenses. I immediately spoke to the concerned doctor in Vellore. He stated that he knew about the financial condition of the father but the amount indicated was the minimum cost involved. We took our decision immediately. On the same day, money was transferred to Vellore. Those who came to request me for helping the boy never expected this. Within the next couple of days, the surgery was done. The news came on the same evening. All our efforts had failed. The boy did not survive the surgery. Though we could not save the life of the boy, I found that the whole incident gave a new hope and new determination to the employees of the Hindustan Copper which was just emerging out of a possible disaster. Every employee talked about it and they openly expressed their feelings that if, despite the most difficult financial conditions, the company could stand by its employees at times of need, they must give their best to ensure that the company could come out of sickness. No sermon or communication could have achieved the level of trust between the employees and the management that this single decision taken out of pure compassion succeeded in achieving.

The second incident happened sometime after the first. A middle-level officer from a unit came and met me at Kolkata. He was suffering from cancer. I was surprised that for his treatment he came to Kolkata instead of going to Delhi or Mumbai which had better facilities for treatment of the disease. He said that he wanted to get treatment in Kolkata because he would like to live with his close relatives who were in Kolkata. At that stage, he wanted moral support more than anything else. After a couple of weeks, the person came to me again. He looked embarrassed and a little sad. I discovered that he was not finding it very comfortable to live with his relatives as it involved a stay for a long time. I could understand the situation. I called the personnel head and enquired if any quarter was vacant and, if so, whether the same could be given to the concerned officer. The personnel head said that a quarter was vacant, but the concerned officer could not be allowed to occupy the quarter because the rule was very specific and a quarter in Kolkata could be given to only such officers who were transferred from other units and posted in Kolkata. If the quarter is given to a person for medical treatment, many such cases will come up and it would be impossible to handle such claims. I told the concerned officer that if the rules could not satisfy a person's genuine needs, then the rules should better be changed. In case quarters are vacant, let the rules provide clearly that it could be given to any person who may be requiring prolonged medical treatment due to a serious disease. The next day, our people got some furniture on rent, the flat was made ready and the concerned person shifted to the quarter. He stayed in Kolkata for the next three months, did the follow-up treatment in Jaipur and got completely cured. He needed mental support along with medical treatment and his company could offer it to him because the company believed that it belonged to him as well.

When I travelled to the mines and plants of Hindustan Copper in mid-January 2004, I found visible changes on the faces of the employees. Most of them looked more determined. They started believing that the company would turnaround and much of the uncertainties that they were facing were visibly over. I never expected that hope would come back to their lives so fast. The workers in Khetri, in particular, were doing miracles. The mine, which had been manned by 1,100 people earlier, was now being operated by just about 400 workers and supervisors and the production was higher than what it used to be in the past. In Malanjkhand, the junior officers were taking the lead in modifying the mine plans so that maximum output could be generated from the mines. They convinced the management that they could do without investing in underground operations, as was proposed earlier. They suggested that with better mine planning and overburden removal, they could increase production to a substantial extent. In Ghatsila, the smelters were shut down, but it did not affect the morale of the employees because they understood the logic. They worked on keeping the smelter ready for eventual use when the Khetri smelter was to be shut down for periodical maintenance work. But despite efforts and all initiatives, in January 2004, the company posted a marginal loss of ₹2.3 million. This was the last time that the company had incurred loss in a month in the history of Hindustan Copper thereafter. The employees of the company, which was incurring a loss of ₹100 billion per month (pm) on an average till November 2003, were visibly shocked to post a loss of only ₹2.3 million in the month of January 2004. They now became confident that the company was going to be turned around. They started believing that it was in their destiny that their company would be a profit-making organisation. Truly, there was no looking back. January 2004 was an aberration, which was never ever repeated later.

One day, while in Malanjkhand, a small group of officers wanted to meet me. I made it a practice to meet the officers, whichever unit I used to visit. Although I had an interactive meeting with the officers the previous evening, I agreed to meet them again as I guessed they had something important to share. It turned out to be correct. Copper ore is beneficiated through a process of crushing, grinding, floatation etc. which is costly because it consumes not only a lot of power but also a variety of consumables. The officers said that it should be possible to find a process of separating copper from ore by using live bacteria. For sulphite ore, the process called leaching was never successful through the organic method, that is, by use of live bacteria. The officers said that they would be happy if they were given an opportunity to work on this thought. I was extremely happy, not only because the idea was new but also because the people of the company, who only a couple of months back were mentally preparing for a disaster, were now looking into the future, becoming innovative and were now being zealous about voluntarily offering their best. The trend was encouraging. I gave them a free hand to go and interact with any metallurgical laboratory they wanted to visit and to carry out any amount of study they would like to do. Despite my busy schedule, I always tried to keep track of the work of those young executives. They failed to make any headway in the initial months and returned empty handed from at least three metallurgical laboratories before they finally made a breakthrough. In a chemical laboratory in Kolkata, a bacteria was developed to carry out the bioleaching of copper through a natural process which would separate copper from ore without involving use of either power or consumables. Finally, in Malanjkhand, a pilot project was set up. The Department of Science and Technology, Government of India, endorsed the move. This further strengthened my belief that human initiative and innovativeness were the best assets a company could have.

....

Companies prosper and perish depending on how the human resources are nurtured in the company.

....

8

STRUGGLE CONTINUES

One of the major elements of cost for Hindustan Copper had been the huge interest burden on the company. It was natural because the company's accumulated losses had been ₹7 billion, a part of which was financed by debts secured from different sources. The banks charged more than 14 per cent interest for drawing cash out of the credit limit fixed for the company. There were unsecured bonds which fetched interest ranging from 14.5 per cent to 15.75 per cent. The company could not repay the amount even after the bonds had matured, but continued to pay interest to the bondholders. It was found that without substantially reducing the interest burden, it would be impossible to stabilise the financial position of the company. As a first step in January itself, I met the top officials of the State Bank of India in Mumbai. My request to them was to reduce the rate of interest and not to insist on government guarantee for the cash credit limit. A government guarantee was not only difficult to secure but it had its own cost as well. The managing director (MD) of State Bank of India was visibly surprised to note that after 90 months, the company had posted profit in December 2003. He promised that both the requests would be acceded to if the company could sustain its profitability.

I had no reason to disagree because I knew that the company had already been put on the right course. In March 2004, I telephoned him and told that I had done my part of the job. He did not take time to reply that he would keep his commitment. The interest rates were slashed from more than 14 per cent to 10.35 per cent. The government guarantee was not insisted upon. As a lead bank, the State Bank of India persuaded all other banks in the consortium to follow them. The company got a huge relief.

Then, I decided to write a letter to the individual bondholders. They were told that the interest rates had already been much lower than what was committed to them at the time of issuing the bonds. They should agree to accept eight per cent rate of interest on the principal amount, and if they disagreed, the company would pay their money back whenever possible. Though the bonds were unsecured, I did not propose any haircut. Every bondholder wrote back within the next few days that they agreed to accept the reduction of the rate of interest as proposed. Logically, this was not surprising, but this had not been attempted earlier because that was not the practice in the public sector. In fact, when I advised our finance head to send the letter to bondholders, he was shocked, and it took time to convince him. Ultimately, I had to prepare and send the letter myself because I felt that every word in that letter was important. Our efforts paid off. The company got a big relief as the savings on account of interest payment to the bondholders were enormous. Till February 2004, Hindustan Copper had an interest burden of about ₹650 million pa, which due to these efforts came down to about ₹400 million, giving a saving of ₹250 million pa. We had, thus, crossed an important milestone towards a turnaround.

Another issue which kept bothering me was that the gross production of the company had became lower after we decided to stop the import of copper concentrate to run the smelters. Our production was less, but the profitability was high. However, HCL's market share came down, and that became a point of concern for us. At that stage, a god-gifted opportunity came. An international copper conference was held in New Delhi around that time, and I was chairing a session in which the main speakers were from the Corporación Nacional del Cobre de Chile (CODELCO) and BHP Billiton. After their lectures were over, I was to propose the vote of thanks and formally conclude the session, but I decided to move a step forward. I not only summarised the speeches but also drew conclusions from their deliberations that were relevant for the Indian copper industry. Now I feel that my analysis was possibly impressive because immediately after the session was over, I was surrounded by a group of people. Two of them in particular insisted that they would like to come to Kolkata to have separate meetings with me. One of them happened to be the MD of a company in Gujarat which was manufacturing copper cathodes only from scrap and not from copper concentrates. His company was not doing well because the import price of copper scrap was higher than what his operational economics could accommodate. After a few days, the gentleman landed in Kolkata and shared with me the issues he had been confronting. We discussed the scope of cooperation between the two companies, and I gave him certain suggestions on initiatives that he could take to source copper scrap from the Indian domestic market.

Immediately after the meeting, I took a stock of copper reverts that we had in our units in Khetri and Ghatsila. Our smelters were not very efficient, and in the process of smelting, the smelters used to generate a good quantity of copper reverts which were basically semi-melted copper concentrates with good copper content which could not be converted into cathodes because of sub-optimum operations of smelters. The generation of reverts is the product of inefficient smelter operations, which in our type of smelters was high. Reverts could either be recharged into smelters along with copper concentrates or sold out to other copper manufacturers. However, the manufacturers of primary copper could use only a small percentage of reverts to blend with copper concentrate to produce cathodes. Most of the time, since there were only two buyers in the Indian copper market, we were not getting reasonable price for reverts. Blending reverts with concentrate in our own smelters was also not ideal because we could not reprocess as much revert as we generated, and further, it involved cost because the reprocessing of reverts required the entire process of conversion to be repeated involving the same cost as was required for converting copper concentrates into cathodes. As a result of all these factors, Hindustan Copper had accumulated a large quantity of reverts in the plant. I had gathered from my discussion with the MD of the Gujarat-based company that his plant in Bharuch, Gujarat, was the only plant in India which could produce 100 per cent copper from reverts because it was a scrap-based manufacturing facility. After a week or so, I requested the MD of the plant to meet me in Mumbai. His company did not have sufficient money to buy our reverts, but it could take our reverts on tolling basis and give us back copper cathodes by charging tolling fees. We were getting a meagre price of ₹50,000 to ₹60,000 per tonne of copper by selling reverts. The proposal, which came from the gentleman, had been that his company would charge ₹38,000, which would cover the tolling fees and the cost of collection of material from our plants and giving it back to our wire rod manufacturing plant in Taloja. The process loss was on his account. The simple calculation was that by spending ₹38,000 per tonne, we could sell finished copper at ₹1,50,000 per tonne making a clear margin of ₹1,12,000, which was 100 per cent more than the price at which we could sell reverts, and that too in small quantities. The offer was exciting, but then I wanted to check up the cost figures once again.

I immediately talked to the cost accountants in Kolkata who were quite excited about the offer. I gave my consent in the meeting itself. Within next two to three days, the deal was finalised. The only snag had been that they were not ready to give us a bank guarantee before taking delivery of reverts from our plant. They could only give us post-dated cheques (PDCs). It was something, which was never done in a public sector company. We sat together and we decided to break the norm once again. The head of our marketing said that it would be his responsibility to do the day-to-day reconciliation between the material delivered and material received, and that he would do it meticulously so as to ensure that the company did not suffer because of any default on the part of the other party. I had seen the MD of the Gujarat-based company only thrice before and did not have any knowledge about him, but I remembered his face and told myself that he could be trusted. The deal was in any case vital for Hindustan Copper and in no way we could ignore it. Within the next couple of days, the party mobilised trucks and dozers in Khetri. They first lifted the material which was visible and readily available. Within the next few days, they started digging all around and discovered a huge quantity of reverts which had been sunk under the ground for over 30 years. The material which ultimately was being collected, existed nowhere in our records. The process continued for month after month and finished copper cathodes started flowing into our plant in Taloja. The quality of cathodes was good. The issue of lower volume of sales which was bothering us in Hindustan Copper was resolved. This single deal gave us volume, revenue and profit beyond our expectation. The head of marketing did his job so meticulously that we never spent any anxious moment due to the fact that the material was delivered on PDCs and not on bank guarantee. Within a few months, the Khetri plant got a different look. The uneven land around the smelter became plain and dressed. With some sprinkling of water it became green. It looked as if nature was rewarding us for our initiative to convert our past inefficiencies into profits by grabbing an opportunity, which was provided by destiny. The other gentleman who had met me in the conference also came to Kolkata to meet me a week later. He was from a large copper manufacturing company in India. Sometime later, we developed a tie-up with his company on the basis of the discussion that we had with him in Kolkata to do tolling of his copper cathodes in Taloja by using our spare capacity for production of copper rods. It was a win-win deal, and we made a lot of money by using our spare capacity.

Incidentally, when we were finalising the deal with the Gujarat-based company, one of our colleagues came and told me that in the public sector, we should not do anything in haste. In fact, placing an order through negotiation without issuing a tender and obtaining quotations and through a process of tender evaluation is never done, and we may be questioned any time by any of the authorities concerned. It was a valid point, but we were in a hurry. The offer was too good to be deferred. We needed more material to be supplied to the market and obviously more revenue and profit. I asked the colleague what his view about the deal was, apart from the procedures. He said this could possibly be the most attractive deal which the company could have. I asked him why did he think so and how we were right in placing the order to the concerned company. He said that the answers were simple. First, it was the only company in the country which could consume the entire stock of reverts that was available with Hindustan Copper. No other company in the country could have bought the entire stock from us because other copper manufacturing companies were only blending a small quantity of reverts with concentrates in their manufacturing process. Second, no other company in the country could have tolled the material for us. Thirdly, our financial benefit per tonne was more than double in tolling than what we could have derived by selling reverts in the national or international market and that was established through our experience during the past few years. Therefore, even if we would have issued a tender for sale of the entire stock, there could be no taker. If we would have issued a tender for tolling, there could be only one respondent and that was the Gujarat-based company which we had negotiated with. I told the concerned person to write down all these in the form of a speaking note so that the arguments are clear and convincing for anybody who could scrutinise the whole deal in the future. When it was written, the position was so clear that no one had any hesitation in placing the order immediately.

This was a great lesson for us. I have seen in my earlier assignments that many a times in the public sector, decisions are deferred or delayed and opportunities are lost just due to a wrong perception that there could be problems from investigating authorities later. If the intentions are clear, the deals are transparent and proper records are maintained about what was playing in the minds of those who were taking decisions at the relevant time, there would be no scope of any problem. At the most, questions could be asked and replies would obviously be satisfactory. I feel tempted to quote an instance in this connection.

Once, in another company, a turnkey order was placed on a notorious agency which was the lowest bidder. The agency never started their work and kept on building records to be able to show later that it was not their fault that the work was not being started. I was advised by the top man in the company to salvage the situation, though it was not even remotely connected with my area of functioning. I studied the tender documents and found out adequate justifications to terminate the contract. Simultaneously, we invited the second-lowest bidder and offered him the job at the rate of the lowest bidder. The party agreed and the order was given on the same day.

I prepared a speaking order, clearly stating that if we did not place the order on the other party on the same day, the L1 bidder could go to the court and secure a stay order. That would jeopardise and delay the project which was of national importance. Next day, the L1 party went to the court and our advocate produced the speaking order which contained all the reasons for which the contract was terminated and also the grounds for which the L2 party was called and given the order. The court upheld our decision. This goes on to show that if the decisions are taken through application of logical mind, which is unbiased, and the facts are recorded in a transparent manner, then it is possible to take prompt decisions in public sector without carrying around an imaginary fear of being harassed later.

Coming back to Hindustan Copper where exciting developments started taking place one after another, I truly felt that luck had begun tilting in our favour. It was evident when the copper prices had started improving. In fact, the price, which was around US$ 1,800 per tonne, reached about US$ 2,200 per tonne by the middle of 2004. This too helped the turnaround process enormously. Our accountants were told to make a month-to-month analysis and to assess how much of our turnaround was contributed by the higher price of copper, vis-à-vis, the initiatives, which were taken by the management. Initially, the price contribution was 40 per cent, which later went up to 55 per cent. This indicated that even if the copper prices were at the same level of US$ 1,800 per tonne, the management initiatives were adequate to turn the company around. But making a profit of ₹100 million per month, which we attained by the end of 2004, would not have been possible without the support of higher price.

As the prices moved up, the mines became more and more viable. One of the mines in the eastern sector which had been closed down in the past became marginally viable. It was an underground mine. Such mines were generally abandoned after closure because maintenance of underground mines required expenses on ventilation and de-watering, which involved cost. If the mines are closed, such costs are not justified and, therefore, after sometime, the mines are allowed to be flooded.

I had an idea even before taking over as chief executive that, in national interest, the concerned mine should not be abandoned. On my insistence, a small group of separated worker, who had formed a cooperative were assigned the job of de-watering and ventilation. That job continued even when the financial position of the company became more critical. Now that the mine became marginally viable, more separated workers joined the co-operative and proposed that they would run the mines if it was transferred to them. I agreed. It required a series of meetings and persuasions to take the proposal forward. I first spoke to the head of the administrative ministry. He agreed. But it was a difficult task to convince many people at different levels. A section of officials in both the Central and state governments were opposed to it. The Secretary of the administrative ministry overruled the objections of his officers and secured support of the finance ministry, which agreed to the proposal to give away the mine to the workers' co-operative on an annual rent of ₹1. The Chief Minister of the state wanted to understand the proposal fully. I met him personally and explained the scheme. He said that he was not sure whether the workers' co-operative would be capable to operate an underground mine, but was keen to watch the result of such an initiative. He added that if the cooperative failed, nothing new would happen, but if it succeeded, it would be an opportunity to reopen a number of closed mines in the state in various sectors. He was excited with the proposal and supported it fully. He knew that the co-operative was managed by a trade union which was associated with the opponent political party, but ignored that altogether. When I briefed the Secretary about my discussions with the Chief Minister, he was reassured and advised me to talk to the Minister of Mines. I did so and explained the proposal in detail. The minister appeared convinced and told me to get the proposal submitted for consideration. However, after the proposal with recommendations from all concerned was placed to the Honorable Minister, something surprising happened. It never came back from the table of the minister. Till now, I do not formally know what happened, but I was told that the Honorable Minister was not happy about the composition of the co-operative, and doubted its ability to carry out the job of underground mining. The minister had better ideas in mind. But then, soon the elections came and the matter was fully dropped. However, despite this setback, the mine was kept ready for operations. Work on de-watering and ventilation was not suspended. After a few years, a contract mining agency was given the job of operating the mine, and till today Hindustan Copper gets the benefit of the output generated by that mine. At the present prices of copper, some other underground mines which were closed earlier could have also been profitable, but those mines were allowed to be flooded and the shafts were lifted and sold at the price of scrap. Surda was an exception, and I always feel happy when I remember that, even at difficult financial times, we had taken a decision to keep the mine alive, even when it was not active.

As the LME prices of copper started moving up creating favourable conditions for us, we got additional value of our efforts, which in any case was driving the company out of red. The combined outcome of all these had been that during the last four months of 2003–04, Hindustan Copper made a net profit despite a marginal loss in the month of January 2004. However, the financial results of 2003–04 did not show profit because during the first eight months, the company's loss was about ₹700 million. So we ended the year with a net loss of about ₹600 million. Every employee felt it was his/her achievement. Someone at the very top level in the government asked me how sure I was that the company could maintain its profit in future as well. I replied that the company had acquired the habit of making profit and it would be very difficult for the company to go back to loss in future. This assessment was proved right all through. Meanwhile, in March, the government released some funds as promised earlier. We were quick to clear up our LC liabilities with the State Bank of Bikaner and Jaipur. A part of the salary arrears were also cleared. The employees of the company for the first time after seven years got the taste of working in a company which could stand by itself. The message percolated down to every level that the health of the company was directly linked with the fortune of its members. Hindustan Copper became a perfect example of how group motivation can work effectively.

Only two issues kept on haunting us. There was a pressure from the government to again close down the Khetri mine despite the fact that the mine was giving positive margin and the output available from the mine was helping the company in utilising the smelting capacity better. The arguments offered from some quarters of the government had been that a public sector company was obliged to follow the decision taken by its owners and it was mandatory. Continuation of mining operations amounted to violation of the decision taken by the government at its highest level. The top bureaucracy in the ministry was never comfortable with this. Another factor which never died down was the proposal to shift the company's headquarter from Kolkata to Delhi and to sell off the office building. On both these issues, we got unexpected support from the newly appointed Minister of Mines. In a one to one meeting, the Honorable Minister told me that she had discussed the Khetri mines with the officials of Geological Survey of India (GSI) and she was convinced that decision to close the Khetri mines was a wrong decision. She added that the initiative to resume operations in the Khetri mines was an action in the right direction. I got the opportunity to explain the background and informed her of the situation. The impact of the resumption of operations of the Khetri mines on the profitability of the company was clearly visible. She had a measured smile on her face. She asked me why I did not announce this in the press. I said that it never occurred to me. She advised that I should call a press conference and tell the media about the reopening of the Khetri mine. I did exactly what she told me to do. It secured a fair coverage in English dailies, but the vernacular press in Kolkata made it a big story. The message to everybody, including those in bureaucracy and political circles, was very clear. The reopening of the Khetri mine had the blessing of the Honorable Minister who being from an alliance party had carried a special weightage in the decision-making process in the Central Government. The dispute and confusion regarding continuation of the Khetri mine was settled once and for all. The matter never came up again. I still appreciate the shrewd move the minister had taken to our great relief. The support on the second issue also came from the same minister, which gave us some relief, though temporarily. One day she called me and told that the corporate office in Kolkata was excellent. I replied that the outer show was very good but the interior was even better. She said that she meant the interior only. I was surprised because the minister had never visited our office and even if she had paid any visit during my tours, I would have got the information. But to my surprise she said that she had been in the office only a couple of days back. It happened after midnight when she was returning from a political campaign in a district and passing through the road on which the building was situated. She suddenly decided to enter the office and move through the floors. I did not know what to say. The security guards did not disclose the fact, possibly because they did not know who they had allowed to enter the office after midnight, but they could not deny entry to the person having noticed the car with red beacon and security escorts. They possibly decided to remain silent without knowing whether what they did was right or wrong. But then, this surprise visit of the Honorable Minister to the office of the company solved many problems. She publicly announced that Hindustan Copper's office in Kolkata was excellent and no one should think of selling it at all. The message was loud, clear and unequivocal, and all those who were insisting on sale of the building were immediately silenced, though the matter came up again after the elections were over and the new government came in. We were again advised to shift the office to Delhi and sell the building. The pressure was mounting every day. During one of the meetings, it led to a very serious altercation between me and the head of the ministry. The verbal duel became so loud and lasted so long that all the officers present in the meeting left the room in order to avoid being the witness of what could happen thereafter. After about half an hour, when they returned, they found that two of us were just chatting and sipping tea. They were visibly surprised and later enquired as to how the situation was diffused. I must admit that the situation had changed because the arguments were never taken to a personal level. The Secretary of the ministry was extremely objective and he never mixed up official issues with his personal relationship with me and his appreciation for my work. Once the arguments were over and it was clear that we could not agree with each other, he decided to ask for tea for both of us and changed the topic altogether. The matter, however, was not over. Even during the last meeting of Hindustan Copper board, which I attended on 31 March 2005 to announce the net profit and successful turnaround of the company, one of the government representatives on the board produced a letter from the head of the ministry in which he was advised to take up the matter of shifting of company's headquarter to Delhi and sale of office building in Kolkata in the board meeting. I asked him the significance of the note. My pointed question to him had been whether it was an expression of desire by the head of the ministry or it was a Presidential directive which was the only instrument through which the government could advise a public sector company or issue instructions. The gentleman stated that it was definitely not a Presidential directive, but it was a desire and advice of his superior. I told the company Secretary to note down in the minutes that the letter was produced and the contents of the same were taken note of by the board which would be examined separately. The matter ended there. Ultimately, however, the company's headquarter remained in Kolkata and the issue met with its natural death.

The success in 2003–04 gave us confidence that we were going in the right direction. The need of the hour was only to consolidate our position and to strengthen it further. We took up three issues to handle immediately.

> First, to develop a long-term strategy for our mines; second, to make our smelters more efficient; and finally, the third, to involve maximum number of people in the decision-making.

So far as the mines are concerned, some mines were already closed. Even in other mines, the essential development work had remained neglected. This obviously happened due to shortage of funds. Once cash flow improved, such work was taken up on priority and this was done in a planned manner. Production of the Khetri mine exceeded the past level by extending the mine up to Banwas, which was an adjoining deposit of good quality ore. The best thing was to sink a shaft to extract ore from the reserve by treating it as a separate mine. However, this required huge capital expenditure and time for completion of the work. The alternate way was to extend the underground tunnel from Khetri to Banwas and extract ore from the same shaft that was used by Khetri. It only involved dragging of material a little longer from Banwas area to Khetri shaft. With the Banwas material coming, not only did the volume of production increase, the quality of material also improved. In Malanjkhand, the overburden removal work which was pending for a long time was taken up. We were keen not only to award contract at the lowest rate but also to be sure that contract went to a genuine party with capabilities. Overburden removal job in any open cast mine is always a little tricky because it is always susceptible to manipulations. Fortunately, we could get some good parties to negotiate with. During the rate negotiation, our team fought for every rupee at every level of negotiation and could place orders to reliable contractors at reasonably good rate. Whenever these initiatives took place, a lot of suggestions and ideas came from employees at all levels. The unions came forward to participate in the decision-making process of management with their visions, ideas, suggestions and apprehensions. I remember, we were selling a huge quantity of scrap through the usual process of tendering. When we almost decided to sell the scrap to the highest bidder, one senior union office bearer informed that we could get a better rate if we could retender. We agreed and really got a better rate. But that was not enough. We discovered that there existed a scope to invite better participation and take advantage of the higher competition and increasing price which was happening in the market. So there was a third attempt to sell the scrap, and we got almost double the price than what we were going to settle for on the first occasion. The excitement amongst the employees in general and among the unions in particular was visible. A very senior union leader, who was always provided with a company car for visiting remote sites, refused to accept the vehicle and decided to undertake a 12-hour road journey in a bus only to set an example for others. When we were thinking of reintroducing the leave travel benefit, many unions suggested that the benefit be continued to be suspended till the company formally posted profit. Meanwhile, we took advantage of not operating both smelters at a time. We got enough time to completely overhaul and revamp both the smelters one by one. Our smelter was consuming more fuel and generating more slags. These problems were meticulously handled. Even in the mini smelter in Ghatsila, we could achieve a drastic reduction in per tonne smelting cost which the smelter had never achieved in the past.

Another very interesting incident deserves to be mentioned here. Copper slimes were available after the process of refining was over. It contained a good quantity of gold and silver. We had a precious metal recovery plant at Ghatsila for extracting gold and silver from the slimes. However, the plant was not very efficient and it was advantageous for the company to sell the slimes in the global market where the price was determined on the basis of efficient extractions of precious metal from the slimes. This precious metal plant was closed, and the slime was sold in the international market. We got very good price. But when we proceeded to sell the next lot of slimes, we got a much lower rate. We discussed this among ourselves. Our feeling was that we were not getting good rates because, possibly, the customers knew that our precious metal plant was closed and, therefore, we did not have an option to process the slimes on our own. We withheld the sale and made a public announcement that the company had reopened the precious metal plant. Photographs of the reopened plant came in the press. The plant, however, was actually not being operated. The whole publicity was for the specific purpose of conveying to our customers that we had our options available. Soon, thereafter, we issued a fresh notice for sale. The rates were substantially higher. Our customers imagined that we are possibly selling the excess quantity. Our game plan paid off.

9

PROCESS OF STABILISATION

As the company started making more and more profits every month, there was excitement all around. It became a personal achievement for every employee of the company. The employees who had all along been perceived to be unconcerned, unaffected and demotivated were suddenly seen to be sharing the excitement of a turnaround with other colleagues with excited faces and joyous smiles. Most employees kept a record of the sales figures and monthly production figures with so much zeal that it looked like collecting and sharing the scores of a cricket match. In fact, work became play, and because of that, everybody started enjoying the work. In my life, I always maintained an open-door policy for all the stakeholders, including employees and customers. I never felt that an executive should ever have a fixed time period meant for visitors. When I took charge of the company, my policy remained unchanged. Many employees coming from different units for official work in Kolkata office would come to me just to express their happiness about the happenings of the company. They would come with a proud face and return with smiles. I looked at them with great satisfaction. Each smile gave me the reward for all the work we were doing day and night. When employees from distant projects get an opportunity to meet the top officials in the headquarters, they normally bring up some of their problems and seek help from the top management. Meetings many a times turn out to be grievance handling sessions. In Hindustan Copper, there was hardly anyone at this stage who came out with any grievance. They only expressed their happiness and silent gratitude. In most of the companies, in their day-to-day operations or in their inter-personal relations, an inherent flow of tensions exists. This is the greatest de-motivator for the employees and the greatest obstacle for them to unleash their full potential. When I observed that the employees were relaxed, taking their job as play, comfortable with each other, including their superiors, and, above all, happy to come to work, I felt that there was no looking back for the company. Now the company genuinely belonged to every employee. The success of the company and its problems were shared by the employees as their own, and this was a stage when it was easy to stabilise the gains achieved and look forward to further improvements. One day in between, I had an interaction with the Secretary of the Ministry of Mines. He was curious to know as to what really was the secret of achieving such a miraculous turnaround of the company. He was quite impressed with our fund management and indicated that, in the initial stage, this helped us to overcome the immediate problems. I agreed but said that maybe a more important factor was that we allowed our people to work without tensions. All the employees were told to pass on their tensions, if any, to me and to take out work from me. This did happen in reality, and it acted as one of the main factors to make the process of turnaround successful.

As the production, both in mines and in smelters, was getting stabilised, marketing became an issue which required some strategic attention. Western India is the largest market of copper. All copper-manufacturing companies were pushing their products in the western region because of the ease to sell, but with its market share, Hindustan Copper was the smallest supplier in this market and, therefore, could be wiped out any moment if the two other large players pursued a more aggressive marketing strategy. For Hindustan Copper, therefore, it was necessary to establish its presence in other markets, especially in southern and eastern India where the other players did not have significant presence. The sale of material, especially copper cathodes produced from Ghatsila in Eastern India, gave a specific advantage to Hindustan Copper in terms of savings on the cost of logistics. Our team became successful in striking a deal with the Ordinance Factories Board to sell cathodes in the eastern market. We entered into a long-term contract, which gave us the required stability. I personally travelled to Kerala to capture the copper market there. We were happy with the response and found the initiative useful. While taking these efforts, we decided that we would not compromise with quality of our products under any circumstances. Our concern for quality was not only to be translated into our actions, it was also to be known to the customers. In one or two cases, the customers came out with quality complaints. We decided not to argue but to accept the concerns of the customers. We replaced the entire consignment with fresh materials. The customers did not expect our response to this extent. We lost some money but gained huge customer confidence, which slowly became our asset in combating competition from larger players.

Another issue which required our attention was to cut down the cost. In Hindustan Copper, we had always been conservative about the costs. We believed in a culture of modesty, free from pomp and show. However, cost-control at an organisational level needed something more. We had to identify the areas of drainage and plug them one by one. One of the important areas of wastage was the money spent on maintaining the Mosaboni township near Ghatsila. The township covered a large area and used to provide accommodation to the officers and staff of the erstwhile Indian Copper Corporation, which was later nationalised and merged with Hindustan Copper. Between 1997 and 2000, most of the mines in this area were closed, except Surda, which was closed later in 2003. All the employees were also separated through voluntary retirement or natural attrition. However, the entire township continued to belong to Hindustan Copper, which was responsible for supplying power and water as well as maintaining the infrastructure, including sewerage. Many quarters were illegally occupied by government officials and police personnel. The markets still continued to exist and drew free power and water. It was impossible to withdraw the power connection due to possible social and political backlash. We requested the state government to take over the township and relieve us of the responsibilities, but the response of the state government was lukewarm. We now decided that we must take it on priority and persuade the state government to accept the township and its liabilities by adjusting the net value of the assets with the power and water bills, which were due to be paid by the company. We used all the persuasive skills that we had and all contacts that we could arrange. Finally, a meeting was convened in Delhi in which the Chief Minister himself was present. The Secretary of the Ministry of Mines initiated the discussion by passionately presenting our case by providing the historical background and the arguments that we had built up. He explained the benefits that would accrue not only to the company but also to the state. Our persuasions created a virtual ring around the Chief Minister. He decided to close the matter on the same day by relenting to our request. Shortly thereafter, orders were passed to take over the township and adjust our power and water bills against the value of the assets. It was a great relief for Hindustan

Copper. The relief was not only in terms of saving money but also in terms of getting rid of the responsibility to get the quarters vacated from illegal occupants, to control power theft and to maintain the township by allotting funds which had better use elsewhere.

In fact, though the company was making profit and the cash position of the company was quite comfortable, the need for money was enormous. During the long phase of sickness, the company could not carry out certain important developmental work which was important to sustain the production of the company. The accumulated backlog of such work was quite heavy and needed funds to be cleared one by one. One such task was to carry out the mine development work at Malanchkhand, which had already been taken up a few months back, but the project needed more money to clear the overburden of non-ore material which covers the seams of copper ore. In order to have access to the ore body in a mine, the overburdens are first to be removed and seams of ore are to be exposed. In fact, in a mine, if the overburdens are not cleared regularly, it is not possible to maintain the same level of production as the ore body—which is hidden beneath a thick layer of overburden—cannot then be accessed for carrying out the mining activity. Therefore, in order to maintain, if not to increase the level of production, removal of overburden on war footing was essential. The production of Hindustan Copper depended largely on the Malanchkhand mine because two-thirds of the total production of the company came from this unit. Fall in production in Malanchkhand would have directly impacted the performance of the company. We decided to mobilise our available resources for development of the Malanchkhand mine. The work moved very fast and we were hopeful that we would be able to overcome the threat of decline in production in this mine. The team operating in Malanchkhand was showing exemplary dedication. I feel tempted to refer to one incident. Once, production from the Malanchkhand mine suddenly came to a halt because the crusher had developed cracks and the crushing operation had to be suspended immediately. The chief of the plant along with a team of engineers and technicians started personally working in the crusher house to put the plant into commission. They sourced the required equipment on war footing and ultimately, on the third day, they could operate the crusher. For the previous 48 hours, the team had worked day and night. They knew that if the production of the mine could not be continued, the smelter operation would be very seriously affected, impairing the performance of the company as a whole. I knew that the chief of the plant had to attend a wedding of one of his closest relatives for which he had taken leave from me. After the plant was commissioned, I remembered this and asked him when he was going to Delhi to attend the marriage. He said that the marriage had already taken place on the previous day and his wife had gone to attend the ceremony. I asked him why he did not inform me earlier. Possibly I could have released him knowing the importance of the ceremony. He replied that nothing could be more important than this and he was happier being in the plant and completing the repair work on time rather than attending a ceremony in which his direct contribution would not have been significant. I silently appreciated his attitude, more so because I came to know of his decision to stay back instead of accompanying his wife only after I enquired about the ceremony. Such attitude was not isolated, but was widely shared in the company as a whole.

We also had to allocate some funds for the development of Kolihan, where also the output was declining due to backlog in developmental work. Kolihan is an underground mine near Khetri which is so well-constructed that one can enter the mine and go up to certain depth in a battery operated railway coach. The mine has a shaft which goes further down. In a copper mine, both underground and open cast, the veins containing copper need to be exposed continuously by removing the rocks around which hide the veins. If this process is not continued, immediate production is not affected, but the sustainability of the mine is affected badly. We decided to start the development work at Kolihan too so that at least a part of the backlog could be cleared.

While we were trying to consolidate our gains and stabilise the company for its long-term growth and sustainability, one issue remained to be addressed without further delay. That was in respect of giving financial relief to the employees. Over a period of the previous one year, delays in payment of salary had been completely eliminated, but the salary continued to be unrevised for quite some time. For all public sector employees, the last revision of pay was done w.e.f. January 1997. In Hindustan Copper, revision of pay could not be effected because the financial position of the company did not permit it to be done. The employees at every level were showing patience. They knew that if the company prospered, it would bring in prosperity to them too. Now that the company improved its performance and came out of losses, we decided that the pay of employees at every level should get revised. I took the consent of the board and decided to hold discussions with the unions and associations to decide on the structure of pay revisions. The discussions commenced, and it was a different form of negotiation that we started. It was not collective bargaining. In fact, there was no bargaining at all. There were no offers and counter-offers and there were no negotiations on the respective positions. There was only collective analysis and understanding of the situation the company was passing through; there was collective assessment of the future of the company, of its accumulated losses and of its future profitability after the pay revisions were decided on. There was already a two-and-a-half times improvement on value additions per rupee of wages. Value addition now became comparable with the best companies in the country. The workers knew this, but they never wanted to highlight it to support their expectations for pay rise. They had a trust that whatever best could be possible would be offered to them.

Meanwhile, my days in Hindustan Copper were coming to an end. In January 2004 itself, I was selected by the Public Enterprises Selection Board to take over Hindustan Copper as its regular Chairman. My appointment as CMD of the company on officiating basis was earlier approved by the appointment committee of the cabinet without any time limit. I told the ministry that I was not interested in accepting the regular position of Chairman because, after completing the process of the turnaround of the company, I wanted to be relieved to take up another assignment. There was huge resistance from the Secretary of the ministry. I assured him that my decision would not affect the company at all because I was already in charge of the company for unlimited period and I would not quit before my task was accomplished. The Secretary was not convinced. He wrote letters to every level of the government to suggest that I should not be allowed to move out because whatever has been achieved in Hindustan Copper could crumble in my absence. The top leaders of the trade unions wrote letters to the authorities at the highest level to stop me from quitting the company. For me, the compulsions were personal. I had to look after my family, and my children were too young for me to retire from service at the age of 58 in Hindustan Copper. The government wanted to do something personally for me, which I was not very comfortable with, as those were to put me on terms which would be different from the other employees of the company. There was resistance within the government as well to have separate terms for an individual. I had a meeting with the Cabinet Secretary, who was convinced that I should be released from the company at an appropriate time. The release order came in the first week of January 2005, but I decided to continue till the end of the financial year. I wanted not only to get the company turned around but also to formally announce that the company was out of sickness before I could quit. On 31 March 2005, we got our unaudited profit and loss account finalised. We called a board meeting on that day, declared a net profit of ₹520 million for the financial year 2004–05, invited the board members for lunch and walked out of the company at 4 PM. The only task which remained unfinished was the introduction of the new pay scales for the employees of the company, which was done soon after my departure. The only anxiety which the employees and their unions did have was that they should not be given too much at the cost of the company. Therefore, even when the pay revision exercise was given the final shape, they wanted to check and recheck with me, even though I had left the company by then, that such pay revisions would not adversely affect the future of the company and its fund position. When I look back, I found that with such morale, there was no reason for the company to get back to sickness, and actually it did not. The company prospered, and prospered well, creating value for its shareholders. Within 10 years thereafter, Hindustan Copper not only wiped out its past losses, it became a dividend-paying company. In November 2012 and July 2013, when in two spells just 9.5 per cent equity of Hindustan Copper was disinvested, the owners of the company, the Government of India, mopped up around ₹10.67 billion, indicating that the market capitalisation of the company was more than ₹110 billion. This was the same company whose owners were almost going to sell it off at ₹100 million in 2003, as was the indication at that time.

In March 2005, I was invited to be the keynote speaker in the Asian Mining Congress held in Singapore. The company by then had come into limelight and started drawing the attention of the concerned quarters beyond

India. I spoke of the Indian copper industry in general and of the initiatives and visions of Hindustan Copper in particular. I spoke about our advancement in bioleaching, and our vision of extracting copper from a large number of small deposits across Rajasthan in particular by creating mobile beneficiation facilities. I represented a country which in the global copper map had an insignificant presence, but our struggle to improve, to innovate and to prosper attracted the attention of everyone present. Soon after the meeting, I was invited by a global business news channel to give an interview. I was watching on the screen that the price of Hindustan Copper shares of face value of ₹10 each had already crossed ₹70. Investors got back interest in Hindustan Copper. It had got acceptance across the country, and the market was stable because China was going to move in a big way in consuming copper. I was immensely satisfied.

A company, being the sum total of the people who work together for attaining the same objective, is also like a living object. It values love and compassion, care and initiative, and attachment and involvement. It also reciprocates love and expresses gratitude. Many years later, I had an opportunity to visit the company once again. I could touch and feel the warmth which the organisation had retained for me. While I was in the company, I suffered financially. I took incessant physical and mental strain. But when I visited the company, I felt that I had received the greatest reward in my life, which could make any rich person jealous.

10

LESSONS LEARNT

In 2004–05, Hindustan Copper came out from the red after almost a decade of loss-making with a net profit of ₹525 million. The company's manpower was by then stabilised at around 5,700 from an earlier number of about 13,000. The value addition per rupee of wages more than doubled to ₹3.70 from ₹1.50 in a few months' time. The marketing officers in different centres had reopened with minimum manpower in order to ensure that Hindustan Copper was never eliminated from the market as had happened a year-and-a-half back. The global copper prices also registered a considerable increase and reached a level of around $2,300–$2,400 per tonne by then. We were confident that the success achieved by the company would be long-lasting, not only because of the improved price and sustained initiatives but also because of the fact that the new work culture had been fully absorbed by the company from which there was no likely come back.

When I look back, I try to assess the real factors which contributed to such an experience of turning around a perennially sick company. Let me try to briefly summarise them.

First, the process of a turnaround is never a miracle. A turnaround happens due to a large number of small efforts, conceived meticulously and pursued relentlessly. Each of these initiatives is relevant at a particular situation and to a particular company and, therefore, it is not possible to standardise them. A story of the great footballer Pele is relevant in this context. A football coaching school prepared a film in those days containing clippings of different moves of Pele, his dribbling, his passes, his shots, his running, his goals and practically all his actions in the field. When Pele heard of it, he was amused. He said it was a waste of time and money. He explained that none of his moves were pre-planned. They were spontaneous, depending on a particular situation on the ground. As no two situations were similar, no action on the field was imitable. What was important was to understand and analyse the situation and develop the capabilities to handle it. The most important thing, which is critical for achieving an effective turnaround, is to have deep involvement of the management in understanding the situation of the company and taking prompt and speedy actions in many areas at a time for bringing about a clear and definite impact. In fact, the management's determination and desperations to change the situation is the key factor for turnaround. Once Shri Ramakrishna Paramhansa had explained to one of his disciples that if he had to experience God, he would have to be as desperate as one who might be pressed under water and grappling for air to breathe. A similar type of desperation, determination and urgency is necessary on the part of the management of a sick company if it has to be revived from sickness. The management has to react to the existing situation with a deep sense of dissatisfaction, a keen desire to come out of the problem and a high degree of impatience towards any delay. Only then can change and real improvement be brought in. Another factor, which is equally important, is to develop a feeling of ownership about the company and to spread that feeling at every level. When I was asked once by some journalists what was the secret of Hindustan Copper's turnaround, I could only smile and say that all of us felt that our company was our family property and that we had to protect it at any cost.

Second, all actions will have to be based on a strategy. Without a clear strategy in place, the actions could be haphazard and non-cohesive. In Hindustan Copper we decided within the first week of December 2003 that mines were the assets of the company, and smelting was a subsidiary function. We analysed that the strength of Hindustan Copper did lie with its mines. It had its own mines, which its competitors did not have. At the same time, even if the company tried its best, it could never compete with other private copper manufacturers in smelting because they were both modern and shore-based, while Hindustan Copper's smelters were small, old and far from the shore. Therefore, the strategy was to use the smelters only to the extent it was necessary to process the ore available from the mines, and not beyond that. This strategy gave Hindustan Copper an edge over others, and all initiatives for turnaround revolved mainly around this strategy.

Third, in the process of turnaround, fast decision-making is essential. Turnaround is a very demanding process and it does not permit the luxury of a time gap between thinking and decision-making and the decisions and implementations thereof. In the process, initial actions may be less participative. However, when the initial phase of actions is over, the organisation will require the largest possible participation of employees in the process of turnaround and decision-making in order to consolidate the gains to the fullest extent.

Fourth, a high degree of motivation of the employees and their high morale is an essential precondition for the successful turnaround of a company. The leader has to lead from the front and motivate the employees by being close to their heart. Many times, in a turnaround initiative, employees are taken as liabilities and not assets. Many turnaround plans start with reduction in employee cost by separating employees by any means and by forcefully depriving them of their dues. In fact, this has been the general method of preparing a revival plan. My experience has been that it never works. I have personal knowledge about the companies which worked on a single premise of reducing manpower and finally landed up being extinct. In Hindustan Copper, rationalisation of manpower was done but employees were never taken as liabilities. Even at the time of deepest crisis, the company took care of individual needs with compassion, valued the employees as heart of the company and treated unions as partners. All these created a feeling of oneness between the employees and the top management, and I have been privileged to witness its impact in turning around Hindustan Copper.

Fifth, communication played an important role in this turnaround process. Clear and continuous communication is needed with employees, unions, other authorities, banks, financial institutions, the press, the public and society at large in all organisations. However, these are more essential for a sick company in its strenuous journey towards recovery. Sickness of a company is a serious disease and it needs everybody's support to recover. In Hindustan Copper, most of the initiatives produced results because of the support it received from government officials, banks, employees and unions. Their support was total and unconditional and was based on the trust they all reposed on the company's management. Incidentally, many companies communicate freely with the employees only when the company's situation is bad but not when it is prosperous, fearing that it would increase the expectations of the employees. Such selective communications are worse than no communication at all. Once the facts are opened up fully and the information regarding all failures and success are equally shared, a company gets the benefit of communication.

Sixth, a turnaround depends to a large extent on appropriate financial management of the company. Without the arrangement of working capital, without securing proper collections and without curbing the interest burden, it is very difficult for a company to come out of sickness. These are the most difficult tasks for a sick company, but without these efforts, a sick company can never revive. In Hindustan Copper, we got the support of the banks, customers and government at an appropriate time. We could cut down the interest burden substantially and could optimise our cycle time of production and, thus, reduce the need for working capital. All these helped and we were lucky to be able to manage our finances well and in time.

Lastly, I repeat, it must be recognised that luck plays an important role in turnaround, but luck favours only the brave. Courageous decision-making is essential, though caution in exercising courage is necessary. Luck, I have seen many a times, comes to favour when it is most needed. In Hindustan Copper's turnaround process, as illustrated on many occasions, we have experienced this luck factor many a time. The only problem with luck is that it cannot be taken for granted; it comes on its own in its own time.

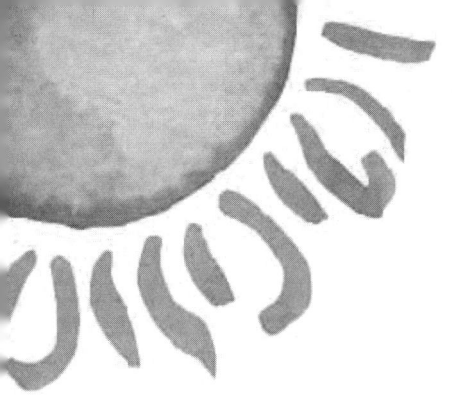

Part II

NMDC—The Sleeping Company Turns into a Giant

1

A GREAT COMPANY

If Hindustan Copper was a sleeping tiger, which on getting an opportunity not only came out of its slumber but also roared back to health and energy, NMDC was not exactly a dormant organisation in that sense. It was already a healthy company, though it was unaware of the potential it possessed. It was happy, making good profits, carrying contended workmen, giving good dividends and providing a peaceful life to all concerned. At a time when the global prices of its products were low, the company had some financial problems, but they were never comparable with those faced by Hindustan Copper. When the prices picked up, an efficient company like NMDC didn't have any problem in taking full advantage of that, and it started tasting prosperity and comfort. When one speaks of a turnaround, it carries a connotation of turning around an entity, which grows from sickness to health, loss to profit and poor performance to good performance. In that sense, the word 'turnaround' is not always used in respect of a healthy company, which already is in profit. But then, turnaround does not only mean recovery; it also means change of direction. It means improvement; it means movement towards growth, diversification and prosperity.

In fact, the turnaround of a sick company or an entity in difficulty is always an easier task than turning around a company that is already profitable. In this context, I am tempted to quote from John Kenneth Galbraith's famous book, namely *Affluent Society*, published in 1958. He writes,

> The poor man has always a precise view of his problem and its remedy: He hasn't enough and he needs more. The rich man can assume or imagine a much greater variety of ills and he will be correspondingly less certain of their remedy .... As with individuals, so with nations.

What Professor Galbraith has mentioned about individuals is applicable to nations and is also equally relevant for business organisations. A sick company has one problem, that is, it is sick, it does not have enough money to take care of its requirements. Therefore, the solution lies in it performing better and making money.

> For an affluent company, the problems are not as well-defined as they are for a sick company. Therefore, the job of turnaround starts right from identification of problems, choosing the directions and reaching a destination which will have multiple dimensions. NMDC was one such company.

NMDC was set up in 1958. It was constituted to investigate, explore and develop the mines with respect to non-fuel minerals in the country. In the early 1960s, it was NMDC which had carried out exploration for copper in Khetri in Rajasthan. Only after the mines were brought to a certain stage of development, they were passed on to a new company called HCL which was formed to specially handle copper mines in India. Ultimately, NMDC was left to do mining of two minerals only, namely iron ore and diamond. NMDC's diamond mine at Panna in Madhya Pradesh was the only operating diamond mine in Asia. NMDC's iron ore mines were spread over a large area in erstwhile MP, and later in Chhattisgarh and Karnataka. The company's affluence was based on the mining of iron ore. The iron ore mined from the Bailadila mine in Chhattisgarh was one of the best in the world in terms of quality. The name Bailadila came from bull's hump. In Bailadia, the hump was the hill full of high quality iron ore, which was easy to extract because of insignificant overburden and high concentration of ore in the mines. The iron ore mines in Donimalai in Karnataka were also the source of fairly good-quality iron ore. Being the largest producer of iron ore in India, NMDC had been a price setter in the market. The prices were also linked with the international prices of iron ore, which was increasing steadily. NMDC's EBITDA kept on improving over a period of time, giving the company a handsome profit.

The Bailadila mines of NMDC were located in Bastar division of Chhattisgarh. Bastar was the largest district at one time, almost equivalent to Kerala in size. Later, Bastar was divided into five districts, namely Dantewada, Bijapur, Narayanpur, Sukma and Kanker. All the districts had a very high percentage of tribal population and huge forest cover. The presence of iron ore in Bailadia range was well-known due to the work done by the geologists during the pre-independence period. But, it was only from the early 1960s that NMDC started working in Bailadia sector to develop the mines. In the midst of forests and tribal population, NMDC set up a very modern mining infrastructure consisting of all the facilities which modern mines could have. Its long conveyors passed through underground tunnels to the loading plant which was fully equipped with stackers, reclaimers, conveyors and wagon loaders. Movement of material through the underground passages proved to be a very environment-friendly system as it did not interfere with the green cover on the surface and did not cause any air or sound pollution. NMDC from the very early days was active in environmental protection. Owing to intensive plantation, the mining site appeared greener from an aircraft than the surrounding areas. Most of the workers working in the mine were drawn from the neighbouring areas. The townships were fully self-contained with good houses, roads, parks, hospitals, schools etc. However, they were far away from the known cities and business centres. The nearest airport at Raipur town, which later became the capital of Chhattisgarh, was about 430 km away. However, the people living in the two townships which were set up for employees working in Bailadila sector were happy and content. Their lives centred around the mines and mining activities. Even their family members took keen interest in the production performance of the mines.

One of the two hospitals used to be run by a very reputed private organisation which could engage specialist doctors in almost all the fields of medicine and surgery. The standard of education imparted in the schools was fairly good with no complaints coming from the employees. The relationship with the local people was good. The doors of hospitals and schools were open for the local population. NMDC was always considered to be a source of support for the society nearby and not a threat. The company was making fairly good profit and the employees were sure that it was capable of meeting their needs. When I joined NMDC on 2 November 2007 and started visiting the mines one by one, I could clearly observe the inner strength of the company.

I soon discovered that the company was one of the lowest cost iron ore producers in the world and that its productivity was that of global standard. The company did not use any contract labour for cost reduction. It only used motivated work force and modern technology to secure high productivity and excellent performance. The unions were supportive, decent and positive. There was hardly any inter-union rivalry. Rather, all unions were part of an apex body called the federation of unions which had office bearers chosen from different unions but maintained strict internal discipline. The senior office bearers were veteran national-level trade union leaders who used to exhibit exemplary magnanimity to move beyond the narrow interest of their own unions and truly represented the interest of the workers. In fact, these senior leaders were more concerned about the health of the company than anything else. They were seen on many occasions admonishing their own members if the demands were excessive and against the interest of the organisation.

Despite all these positive factors, the company was apparently suffering from some sort of stagnation and complacence. It had the same production, same approach to sales, same mines, same manpower, same system of management control, almost same revenue and same profit. There was a sublime feeling of superiority among the mining engineers, who felt that in a mining company, they knew the best, and the top slots were always reserved for them. An excellent mineral processing engineer was not considered to be appointed in the board just because the mining engineering community felt that he was not fit to be the head of production. In fact, he had to wait till my joining the company to become a director and he proved himself to be one of the best. The employees in general were proud of the fact that NMDC belonged to them and they to it. But they also felt that their interests were best protected if they maintained a lowkey presence.

For many years, the company's activities on iron ore mining were confined to mainly Bailadila in Chhattisgarh and Donimalai in Karnataka. The company carried out mineral exploration in a number of mines in different states, but none of them were converted into mining leases because of either litigation or lack of tenacious persuasion on the part of the company. Consequently, the area of operations of the company remained confined to these two sectors only. In other mineral-rich states such as Odisha, Goa and

Jharkhand, NMDC did not have any presence. One important thing needs to be mentioned here. Wherever the company did any work, be it construction of mines, townships or laboratories, the quality of work was excellent, the plans were well thought of and the execution was excellent. Even when a commercial policy was drawn, it was so well-devised that during the recession of 2008–09, I could clearly see these policies withstanding the fluctuation in market and volatility in prices across the world.

When I joined NMDC as its CMD, I had only theoretical knowledge about the company. In the first couple of weeks, after visiting the mines, I developed a fair idea about the operations of the company, the attitude of the people, the technology used and the infrastructure the company had created in the remote areas where the mines were situated. On 15 November 2007, the company stepped into its golden jubilee year. NMDC was set up on 15 November 1958 and 2007–08 was the golden jubilee year for the company. I returned to Hyderabad from a tour in the afternoon and at 6.30 PM, I reached the venue of the programme in a large hotel in Hyderabad. The hotel lawn was full of people, existing and past employees including the previous chairmen and directors and senior officers of the company. I had to inaugurate the golden jubilee celebration and deliver my speech. I was watching the reaction of audience and their deep emotional involvement with the company. I interacted with the past CEOs, directors and others and had the opportunity to listen to their feelings about the company. They were curious to know whether I had seen many of the things they were referring to during my visits to the mines. Most of them had spent their whole lives in the mines and knew the mines like the palm of their hands. The ladies came forward and joined the men in their reminiscence of the glorious times they had spent in the mining township. I was watching their faces and listening to their comments.

After some time, I walked out of the venue, went away to a corner of the hotel and telephoned my wife. I told her that I had come to a great company with huge strength and great expectations of people from me. They expected the company to be nourished, protected and developed by me. I told her that I must meet their expectations and improve myself further to be fit to lead such a glorious company and that destiny had given me an opportunity to serve a great company like NMDC. My job was to ensure that I accepted the responsibility with humility and be competent to do justice to the responsibility bestowed on me. My wife always remembered those words and gave all the support I needed from her to be able to serve NMDC as my only passion.

I have seen many a times in life that destiny plays an important role in a person's life. Often I feel that my destiny dragged me to NMDC. As I look back, I feel that my only job was to do justice to what destiny had offered to me; to embrace the opportunity with humility, with open arms and with all sincerity and to rise up to the standard which was demanded from me. All my initiatives to change the course of the company, to turn it around and to convert it from a local company to a global company were the results of this conviction of mine.

Another factor which I strongly believed in, and I do believe always, is that people are inherently good. I believe in nourishing and nurturing their goodness. Even those whom society thinks are bad have plenty of good qualities in them which come to the surface if they are given a positive environment and proper support. People can take a company to any height if they are properly motivated. They are prepared to sacrifice and can come out of their petty interest if they are convinced that what they are doing is for a greater cause. I developed these feelings strongly when I was in Hindustan Copper. In NMDC too, I had no reasons to think otherwise. The journey, which I had taken in NMDC in early November 2007, ended with my superannuation on

31 December 2011. During these 50 months, I experienced a revolution in the company. New experiences unfolded almost every day. I developed a deep appreciation for the strengths the company had and the determination exhibited by the employees to grab new opportunities, extend into new areas and see things differently.

My story of NMDC is a story of growth and diversification. It's a story of a company elevating itself from strength to strength, expanding itself from one business to multiple businesses, one geographical area to different areas, and all these together contribute to the story of its turnaround.

2

THE JOURNEY BEGINS

Within a few days of my joining NMDC, I was traveling from the state capital Raipur to the Bailadila mines in Bastar. It was a long journey, but a pleasant one. The distance was 430 km. After 300 km, when I crossed the divisional headquarters at Jagdalpur, I experienced a distinct change in scenery. The surroundings became greener, and between the hills and plains, a magnificent landscape began to emerge. There were tall trees on both sides of the road and slowly the tropical forest unfolded itself. After about two hours' drive from Jagdalpur, suddenly the township of Kirandul, the oldest mining township of NMDC in Bastar, became visible. In the midst of hills and forests, Kirandul was a well-laid-out residential colony for NMDC employees. Around Kirandul, some urbanisation had also taken place, which included transport depots, markets, garages, a few supporting industries and so on. Due to the concentration of local population, Kirandul area was brought under a municipality. Beyond the limits of the municipality, there were small settlements of local people, mostly tribals, some of whom were already employed by NMDC in its mines, but many of whom were surviving on farming. NMDC was already supporting the local population through its corporate social responsibility (CSR) initiatives by setting up school buildings and community centres, building roads and bridges and providing health care services to the surrounding villages. All the districts of Bastar division, namely Dantewada, Bijapur, Narayanpur, Sukma and Kanker, had been witnessing Naxalite activities for quite some time. The areas surrounding Kirandul were at the centre of those activities. A few months prior to my joining, some Naxalites entered the mining area and burnt a large part of the conveyor system of the mines. The entire mining area of NMDC was guarded by the Central Industrial Security Force which was fully equipped with all possible arms, ammunitions and gadgets. But the area was too large to be guarded and there was always a risk of Naxalites sneaking into the mines and causing damages.

Before I talk about my first visit to Bailadila, let me describe a little about its history and background. Bailadila falls in Dandakaranya region mythologically known to be the abode of Lord Rama after he was banished from Ayodhya. Ramayana tells us about the natural beauty of the place and how much Lord Rama adored the forest life of that region during the 14 years of his exile. Even today, the beauty of nature to a substantial extent remains undisturbed. During the British rule, an Indian geologist first indicated the presence of high-quality iron ore in the area. During 1955–56, Professor Euemura of the Japanese Steel Mills Association studied the early reports of the Geological Survey of India and informed the Japanese steel mills about the richness of the vast deposits of iron ore in this area. Subsequently, Japanese steel mills had shown interest in the project in 1960. In 1964, NMDC prepared the project report which was followed by construction work. The first iron ore mine and plant came up in Kirandul area in 1968. On 1 January 1948, Bastar had become a part of the Indian Union. In 2000, after Chhattisgarh was separated from Madhya Pradesh, the area became a part of

Chhattisgarh. Bastar is now known for its natural wealth, mineral resources, large tribal population and Maoist insurgency.

The main connecting link of Bailadila with the port of Vishakhapatnam (Vizag) is the Kothavalasa–Kirandul railway line through which most of the iron ore mined by NMDC is transported to Vizag port for export or domestic consumption. This railway line has a track length of 443 km. It passes through 58 tunnels and over 84 major bridges. This is also the highest altitude broad gage line in the world which touches a height of 996 m above sea level. The work on this project was taken up in the early 1960s and the line was commissioned in 1966–67. When I joined, about 15 million tonnes of iron ore used to be transported from Bailadila to Vizag through this route. A major part of it was consumed domestically by the Vizag steel plant and two other privately owned steel plants situated in the west coast. A portion of the materials transported was also exported to Japan from Vizag port under a long-term export contract. A private steel plant had also laid a 267 km long slurry pipeline, which used to carry iron ore fines in slurry form from Kirandul to Vizag for conversion into pellets and transported to the steel plant at Hazira. The total evacuation capacity being limited to about 23–24 million tonnes per annum (MT pa); the production capacity of the Bailadila mines was restricted to that level. In Bailadila, NMDC was operating a number of mines and its total production could easily exceed 30 MT if it had not been restricted by the available logistics for evacuation of material.

In Bailadila, NMDC had two mining colonies, situated at a distance of about eight km from each other. While Kirandul was the older township, the newer one was Bacheli, which also had all the modern facilities required for the employees to enjoy a modest living standard. Bacheli had a good hospital that was run by Apollo Hospitals Ltd and funded by NMDC. After inspecting the mines in

Bacheli, I decided to pay a visit to the hospital. I found that a large number of beds were occupied not by the employees but by the people of nearby villages who were admitted to the hospital. These people were getting not only free treatment for themselves but also free boarding and lodging for their companions. I inquired about every patient from the chief of the medical services, especially about those who were admitted from the nearby villages. I moved from one bed to another and the doctor kept telling me about the disease that the person concerned was suffering from. The doctor's commentary was like this; '[H]e is suffering from acute Jaundice, the next person is suffering from anaemia, this is the man who was brought to the hospital in a very bad condition, he was suffering from malaria. This person is suffering from typhoid', and so it went. There were patients with acute skin disease. There were patients who were brought to the hospital almost in a dying condition, suffering from cholera. After I finished a round of the concerned ward, the doctor followed me. He said, 'Sir, they all are practically suffering from only one disease i.e. poverty'. My feelings were also the same. It was poverty and acute poverty which was the cause of these diseases, and as long as poverty exists to such an extent, there would be no dearth of people who would continue to believe that the barrel of the gun was the only source of power.

While I moved through the mines and the allied facilities, my experience in the hospital continued to haunt me, but soon my mind was shifted to the massive work that was being carried out in the mines. I could see the great care that was being taken to ensure that every ounce of ore was scrapped out of the mining pit with motherly care, lest there be any wastage. I could see that the quality of ore was so good that some of the lumps looked like almost a full piece of iron with no gangue in it. The process of mining in an open cast iron ore mine is not very complex, especially in mines like those in Bailadila, which are open cast mines with hardly any overburden. Therefore, whatever slices were cut out from the mining pit, contained only ore, which only needed to be crushed and screened to be transported to the loading point. But then, all these activities were handled so methodically and so carefully that there was hardly any wastage of ore and hardly any selective mining to leave out some ore bodies behind. I could visualise how sincerely about 3,000 employees were working in that sector, being fully involved in producing and supplying to the nation much needed iron ore required for making steel and building a modern India.

However, on the flipside, I could also observe that despite the intensive CSR work carried out by the company, the wealth of the company did not percolate down to the regions which were still suffering from unemployment, poverty and malnutrition. A mining company cannot provide direct employment to a very large number of people. Even if indirect employments are included, the total number of people who could be employed would not be much as compared to the number of people aspiring for jobs. Mining is, therefore, perceived to be an exploitative industry, which exploits and extracts resources from the womb of Mother Earth and takes them away for wealth creation elsewhere. Despite all the good facilities which the townships of Kirandul and Bacheli did have, they looked like islands of prosperity in the midst of the darkness of poverty. This possibly had been the situation not only in the mining townships but also in other industrial townships set up in the 1950s and the 1960s in India. Beneath the glittering urban settlement in those townships there were dark habitations of local people who were deprived of the benefits of development, which they surely deserved.

Soon after my visit to Bailadila, I went to the Donimalai mine in the district of Bellary in Karnataka. The experience was totally different. In Bastar, the only mining company which was operating was NMDC. In Bellary, a large number of mines were in operation. Due to heavy truck movements, the roads were broken and in some patches they did not exist at all. There was dust all around. The fields on both sides of the road were covered by red dust. Some of the mines which could be seen were clear examples of how unscientifically the mining operations could be carried out. There was relief only after we reached Donimalai township and visited the Donimalai mine. The township was well-built and well-maintained, and so were the mines. The system of mining was similar. The quality of ore was, however, marginally inferior. The Donimalai mine at that time was considered to be almost exhausted and a new mine on the top of a nearby hill called Kumaraswamy was being planned to be developed to substitute Donimalai. The hills were green and the environment was well protected. The mines, however, contained large dumps of waste rocks, mainly banded hematite jasper (BHJ) and banded hematite quartzite (BHQ), which had low iron content and, therefore, were treated as waste. The people of the surrounding area were not as poor as those in Bastar. It was not only because a large number of mines were operating in that area but also because many people were engaged in allied work. Some were even involved in petty extortions. The rupture of environment, which was evidently taking place in Bellary, was the result of the reckless mining of small miners who were driven by greed and greed alone. This clearly percolated down to the community as well. Many people felt that they also did have a share in the easy money earned by the miners by exporting iron ore to China at about 300 per cent to 400 per cent profit. The display of wealth by miners was so vulgar that Bellary was not only degraded environmentally but also morally. The difference between the social situation of Donimalai and that of Bailadila was distinctly visible, and the environmental degradation which was taking place in Bellary was so evident that it was not surprising that ultimately the highest court of the country had intervened and stopped the mining operations in the region at a later stage. The only mine which was exempted and allowed to continue functioning was the Donimalai mine of NMDC. This, however, will be discussed at a greater length later.

After Donimalai, my next target was Panna. Panna is located in Madhya Pradesh near Khajuraho. The mine is very close to the Panna National Forest and Wildlife Sanctuary. By the time I joined NMDC, the operations of the Panna mine had been closed. The mine was in operation for many years prior to that. However, it was suddenly discovered by some executives of the concerned department that the mine was actually located inside the wildlife sanctuary and not outside it, as was earlier presumed. Therefore, suddenly, a ban came and the mining operations came to a halt. The Panna diamond mine had been the only diamond mine in Asia and had been a place of pride for NMDC. After the closure of the mine, the matter was taken up before the Green Bench of Supreme Court, where any mining operations in a wildlife sanctuary were vehemently objected to. Our officers said that it was already a lost case and there was no option but to close down the mine permanently and shift the people to other units. However, I felt that we should fight up to the end. I decided to visit Panna and see for myself the worth of the mine. Financially, Panna had never given NMDC very good profit. It was always profitable, but the percentage of profit was never comparable to that which was earned from iron ore mining. However, Panna had its own significance. In a remote area of Madhya Pradesh, Panna was the most important economic activity that existed. It was not only employing several hundred people but was extending valuable facilities to the nearby villages from its own resources. Above all, Panna was producing about 70,000 to 80,000 carats of diamond and the closure of that mine would deprive the country of the natural resources which the mine could make available to the nation. To me, closing a mine, knowing that it had deposits which were valuable for the nation, was unacceptable. The facilities available in Panna were also reasonably well-built and should not have been abandoned if there was any way to save them. After visiting Panna, I was convinced that the mine, which was being operated for many years without impacting the wildlife sanctuary, could not have caused any harm to the environment and wildlife. The best thing in the interest of the country should be to completely extract every piece of diamond available in the mine as fast as possible and then to close down the establishment. Having been convinced that this should be the right approach, I decided to consult a lawyer who was earlier associated with the company and known to be an honest and intelligent person. We had a long meeting in which we tried to understand and analyse every logic which had been forwarded for closure of the mines. We decided to take the matter beyond the legal issues and to focus on the national interest and, thus, convince the Honorable Court that whatever we wanted to do was in the best interest of the nation including the interest of environment and protection of wildlife. During the next hearing, our lawyer took this approach. He lifted the entire discussions above the level of legal arguments and spoke of the points we discussed. This worked. I am told that, in the court somebody observed, though jokingly, that NMDC was such a competent company that may be the tigers would be safer in the hands of NMDC. The verdict came in our favour and the mine was reopened. There was celebration everywhere. When the formal reopening of the mine took place sometime later, the Honorable Minister asked me how much of the profit made by the Panna mines should be allotted for community work around the mining region. I said 100 per cent. The minister did not believe it. He asked me the question once again. Then I explained that for us in NMDC, reopening of the mines was more important.

The profit generated were not really as much as we could make from other operations and if 100 per cent profit was shared with the community, the entire community would morally and sentimentally own the establishment, which would make the mining operations more sustainable in the area than otherwise.

The Honorable Minister saw the point. During his public speech, he, however, restrained himself from using the word 100 per cent. He said that the company would look after the interest of the community and as much profit of the unit as required would be allotted for the purpose.

After hectic travels for a couple of weeks, when I returned to Hyderabad, I decided to pay a visit to the research and development centre of the company located in another end of the city. The centre was one of the best laboratories on mineral processing in the country. It was created in 1970 and was catering not only to NMDC but also to a large number of other companies. It was involved in various activities relating to mineral processing, flow sheet development, mineralogical studies and project development. It had been declared as a Centre of Excellence in the field of mineral processing by the expert group of the United Nations Industrial Development Organization (UNIDO). It had secured a number of patents, the most important of which was the productive use of kimberlite waste found in diamond mines. It had also developed an appropriate technology in association with Russian experts for the production of carbon-free sponge iron powder and had set up a pilot plant on this. Despite laudable achievements of the past, the research and development laboratory appeared to have been suffering from some sort of stagnation and complacency.

Back in office, I summarised my conclusions as follows: NMDC was a strong company. Its strengths had already been absorbed by the company in its core features and, therefore, it was in right health to grow further and prosper. Second, the company's operations were confined within the same area for many years and, despite its efforts, it could not acquire any new mines. Even the iron ore mines where NMDC had carried out explorations were denied to the company on some grounds or the other. The excellent capability of the company was not practically being used by the country to its full potential. While, with the boom in iron ore market, hundreds and thousands of mining leases were being distributed to small private players who had neither the resources nor the technology to carry out scientific mining, NMDC's claim was being largely ignored. The company was not being able to make much inroad into securing new leases to expand its iron ore mining activities. It had only succeeded in obtaining a mining lease of a limestone mine in Himachal, a magnesite mine in Jammu and two small underground coal mines in Madhya Pradesh. These, however, were only peripheral activities for NMDC and not a part of its core business. Third, the company was extremely shy to move beyond mining and get into value addition. Years back, the company was given 1,000 acres of land in Bastar in which the company decided at some stage to set up a sponge iron plant, which did not get encouragement from any quarter. I felt that the company should move forward both on acquisition of new mines and on value addition if it had to sustain, grow and remain strong and relevant. Thus, began the turnaround process of NMDC.

3

NMDC EXPANDS INTO STEEL MAKING

In the earlier part of the book, I had stated that in any turnaround, luck plays an important role. NMDC was also not an exception to this statement. Within a couple of months of my joining, it was becoming clear that NMDC would get the status of a Navratna company. This not only gives more powers to the board but also gives the company the moral right to take certain independent decisions without requiring the approval of the owners, that is, the Government of India. In other words, Navratna status establishes a psychological contract between company's management and concerned administrative ministry that the company should enjoy a fair degree of independence without interference from the ministry in addition to enjoying certain formal powers, which could be availed by its board. In February 2008, the Honorable Minister in charge of steel decided to pay a visit to the Bailadila mines for the first time. I joined him. He was taken to every part of the mines. He was visibly proud to see the company's activities and achievements. In a public meeting arranged in Bailadila, the Honorable Minister made an announcement regarding NMDC being given the Navratna status. Suddenly the company came into the limelight at the national level. A large number of journalists had flown to cover the visit of the minister. To all of them, Bailadila was an unknown territory. I wanted to take this opportunity to make the strengths of NMDC publicly known, which was essential to make a headway on the plans of development and diversification, which I had in my mind. I addressed from the public platform, and made the following appeal to the media,

> Dear friends, when you go back to Delhi please tell our countrymen that we, 3000 brethren of theirs are working in this remotest part of our country, day and night to extract iron ore from the womb of mother earth only to make their lives prosperous and the country stronger. When we work we know that every bit of our work goes ultimately to making steel and building the nation.

I added that 'each employee of NMDC is a soldier fighting against all odds in one of the most inhospitable regions of the country and silently and incessantly working for his or her nation to give it the strength of steel'. The meeting ended. Everybody returned, but the employees of the company living in the sleepy towns of Bacheli and Kirandul appeared to have woken up to an understanding which was not new but was most emphatically spelt out this time, that what they were doing was relevant not only for their families and their company but also for the nation. They were, in fact, doing a national service by providing inputs for making steel, which in turn built the infrastructure of the nation for its industrialisation and growth. They knew the perspective of every bit of work they were doing and realised that each one of them had been a soldier for nation-building in the truest sense.

Despite possessing the best mines of the country, good technology and work culture and good performance, a threat was looming large on the company. For many years, NMDC was the only mining company operating in Bailadila sector and all the mines in the region were handled by the company. Right from the late 1950s, NMDC's presence in Bastar and its natural claim on the Bailadila mines had never been questioned despite changes in the government in the Centre and the state, and bifurcation of the state and creation of a new state called Chhattisgarh. The Naxalite insurgencies acquired new grounds and new strengths in the area, but NMDC remained an integral part of the mineral-rich area of Bailadila. This position was first threatened in 2006 when the state government decided that the mines in that area would be given to those companies which would agree to set up steel plants in the region for value addition and employment generation. As a merchant mining company, NMDC never felt the need for doing anything beyond mining. It followed the traditions of large mining companies of the world, which had mining as their area of core competence, and they never felt that they should dilute their position by getting into any other activities beyond their core competence. The situation in India was, however, different. India was neither Australia nor even Brazil with high landmass and thin population. A highly populated country like India needed employment and income generation in a big way. The states which had mineral resources wanted to leverage their mineral wealth for getting large industries set up for generation of income and wealth. The state of Chhattisgarh, which had the best iron ore reserves of the country in its possession, was no exception. It decided to have agreements with two privately owned steelmakers and extracted a promise from them to set up steel plants in Bastar in exchange of allocating iron ore mines to them. All iron ore deposits in Bailadila were either with NMDC or happened to be the natural extensions of NMDC mines. The state government terminated the lease of NMDC in respect of one of the deposits and gave it to a company in the private sector. Another deposit, which had been practically an extension of NMDC's mines in Bailadila, was denied to NMDC and given to another company, which also promised to set up a steel plant in the state. NMDC's position was threatened in Chhattisgarh for the first time in its existence in the state. We decided to put up a fight. The Government of India, whose permission was required for NMDC to file a suit against another government body or government department, cleared NMDC's proposal to start litigations against the state government for granting prospecting lease to two private companies as well as the Central Government for giving prior approval to the state in this regard. The relationship between NMDC and the state government, which had been cordial for all these years, now became bitter. The state government viewed these initiatives as anti-development and blamed NMDC for neither doing any value addition on their own nor allowing others to do so. One of the private companies moved fast to acquire land to set up the steel plant. The state government gave them full support.

By the time I joined NMDC, the wide trust gap between the state administration and NMDC was clearly visible. The relationship between the state government and the administrative ministry also did not appear to be very cordial on the ground that NMDC filed the court case against the state's decision under the instructions of the administrative ministry. Meanwhile, certain local political groups started agitation against land acquisition by one of the private companies, which had been promised an iron ore deposit as a part of the memorandum of understanding (MoU) signed between the state and the concerned company. The state government even at the highest level gave an impression that the agitation was instigated by NMDC and its unions. When the Cabinet Minister in charge of steel visited NMDC's mines in Bailadila, the state government with apparent reluctance nominated a junior minister to share the dais with the Central Minister. In talks and gestures, the state government was determined to air their unhappiness as much as possible. I felt, as the Chairman of the company, that what was happening was not hygienic for the company. For any industry especially for a mining industry which is involved in extraction of major minerals such as iron ore, state support is essential not only for the existing operations but also for future growth and sustenance. It was, I felt, unwise on the part of NMDC to allow the relations to deteriorate, for which the company, not the government would be the ultimate loser. When I spoke to some of the senior district officers, I found them candid in their arguments. They said that despite violent resistance from the tribal people, the state government had handed over 1,000 acres of land to NMDC long back to set up a value addition project, but nothing practically happened thereafter. The fact is that, NMDC was never keen to set up a conventional steel plant. The company decided to acquire and use a Russian technology called ROMELT for converting the iron ore fines into liquid iron or pig iron. Making steel in a small scale by using this was also planned. ROMELT technology was developed in Russia when it was a part of the Union of Soviet Socialist Republics (USSR). The technology was being developed by three constituent states of USSR. However, after the disintegration of USSR, the project ran into difficulties. There were some efforts to retrieve the project, but the pilot plant which was set up did not achieve success. Therefore, there was no meaning in acquiring an unproven technology and using it to set up a 600,000 tonne capacity plant in the land acquired by NMDC. The company then came out with an alternate idea to set up a sponge iron plant of same capacity. The state government did not view these efforts with enthusiasm. They felt that NMDC was not sincere in their plans to do value additions, and what they were practically planning to do was only to waste time and make a token investment to satisfy the government's expectation that there would be value addition and employment generation in the state through industrialisation. When I took stock of the situation, I found that at long last tenders for setting up the sponge iron plant had already been floated and responses had started coming in. However, about six months before my joining NMDC, an MoU had been signed by NMDC, Steel Authority of India Limited (SAIL) and Rashtriya Ispat Nigam Limited (RINL) to set up a steel plant of three million tonnes capacity in Bastar, for which separate land was to be located and acquired. As per the plan, NMDC would provide iron ore and SAIL and RINL would provide the technical expertise and supervision for construction of the plant. After signing the MoU, not a single meeting amongst the parties involved had taken place and there was no movement on the issue at all.

I coolly assessed the situation. The expectation of the state government that there should be value addition in the state based on the mineral wealth available in abundance in the state was a reasonable one. Similarly, there was genuine reason to believe that NMDC was not at fault in sticking to its area of core competence and not stepping out of the same for vertical integration of its business in which it did not have any expertise. But then, it was necessary to resolve the differences. NMDC's mines were due for renewal within the next few years and its strained relationship with the government could come in the way of the very survival of the company. Incidentally, the state of Chhattisgarh had a large number of sponge iron units and the state knew that they were creating more pollution and less employment. The state government was in no mood to accept that NMDC would be setting up another sponge iron plant despite the huge mineral wealth available at its disposal.

I had earlier met the Chief Minister of the state, but then I decided to meet him again and understand his mind on this issue. I observed that his attitude towards NMDC was quite stiff. He felt that NMDC, which was granted most of the iron ore leases in the state, had acted against the interest of the state by going to court against the state's decision. He no longer needed NMDC to do any value addition in the state because the state had already allotted two iron ore mines to two private companies, on the understanding that both the companies would set up integrated steel plants and use the entire production available from those mines for steel-making in the state. One of the companies had promised to set up a steel plant of 5 MT capacity pa and the other company promised a steel plant of 3 MT capacity. The Chief Secretary told me separately that before the private companies were given the mineral concession, the erstwhile Chairman of NMDC was clearly asked whether NMDC was prepared to set up a steel plant and if so, there would be no agreement with any private company. It was claimed that the MoUs with private companies were signed only after the then Chairman of NMDC had expressed his inability to get into steel-making because NMDC was primarily a mining company and mining had been its core area of competence. I had a fair understanding of the respective positions of the state government and NMDC, and felt that the situation could not be allowed to drift in the way it was and that it had become necessary to act immediately. Within a few days, the Honorable Minister of Steel had convened a meeting in Delhi to assess the progress on setting up a joint venture steel plant in Chhattisgarh for which an MoU had already existed among SAIL, RINL and NMDC. During the meeting, SAIL and RINL stated that they would be interested in being a part of the project provided that the iron ore was supplied by NMDC at cost price or transfer price, which would be very close to the cost. I found that no interest of NMDC would be served if NMDC provided iron ore at cost price, invested in the plant and derived 1/3rd of the profit from the joint venture company. It will be perceived by everybody as a joint venture plant and not a plant of NMDC. It was necessary to convey a message to everybody—the government and the people—that it was NMDC, which not only extracted ore but also added value to it by making steel. The most important point against this project was that despite signing of an MoU the project had not moved an inch because nobody knew who was accountable for it. Even eight months after signing the MoU, these three companies had never met with each other and discussed the project at any level. When the ministry decided to review the project for the first time after eight months of signing the MoU the project was remembered and talks on the project started again from zero level. I remembered a proverb; 'a mother of many sons is always unattended.' I decided to tell the Honorable Minister in the meeting what I was deeply thinking of for the previous few weeks. I shared with him my thinking and told him that the joint venture concept should be scrapped not only because it was not going to do any good to NMDC but because a greenfield project for setting up an integrated steel plant could never be executed jointly because the project would be nobody's baby and structurally the process of decision-making would be so slow that it would be impossible to execute. I stated very confidently that NMDC would set up a steel plant on its own and the plant would come up within the minimum possible time. Many officials present in the meeting were skeptical. They said that NMDC did not have the expertise to make steel. I took the names of some of the private sector steel-makers and told them that many of the so called private sector steel makers did not have any background of making steel. They only knew how to hire the right people, who could work in the project and give them the right direction. NMDC, I said, would hire right people, take guidance from right quarters, consult the right advisors and set up the steel plant on its own. I was emphatic, emotional and candid. Everybody appeared convinced. I knew that it was a challenge. I knew that it was a commitment that was not easy to fulfill, but I also knew that we were already late and unless, we, from NMDC set up a steel plant on our own, we would not convey the right message to the state government. We would also not be in a position to meet the aspirations of the people in Bastar who had always seen us involved in extraction of wealth in Bastar and not creation of value in the region, which would offer employment, earnings and better life for the people. I knew that the steel plant would be the biggest guarantee for NMDC to continue its operations in Bailadila with the support of the state and the community. I also felt that by investing in steel plant, the position of NMDC as a merchant miner should not be allowed to be diluted.

I decided to persuade the state government to allocate a fresh mine to NMDC, which would be the captive mine for the steel plant. The idea was not to disturb NMDC's existing merchant mining operations in Bailadila and to ensure that the customers of NMDC in India or overseas continued to receive the same quantity of iron ore if not more from NMDC even after the steel plant came up. NMDC already had 1,000 acres of land in a place called Nagarnar, which was very close to Jagdalpur, the divisional headquarter of Bastar. However, the area was insufficient for setting up a steel plant of 3 MT capacity. We needed additional 800 acres of land. Considering the resistance which the company had faced earlier while acquiring the land already in possession, securing further land in the same area was evidently not an easy task.

After returning from the meeting, I explained the position to my colleagues. Initially they were not very enthusiastic about the idea of going into steel-making. I explained the situation to them analytically, clarified the logic and of course presented my vision. I told them that I had full faith on the strength of the company and I was sure that unless the company was diversified through horizontal and vertical integration, slowly, but surely, the company would lose its vitality and relevance. I explained to them how the company had lost two mines in Bailadila itself, one of which was already under company's lease and surprisingly, the Ministry of Mines, Government of India, had supported the state government and not NMDC, despite the fact that NMDC had been a Government of India enterprise and that it was an apparent injustice to NMDC that a mining lease was prematurely terminated and the deposit handed over to another company. I continued to explain that despite such an injustice, the company's reaction was limited to filing of a court case and earning the displeasure of the state, none of which was going to help a commercial organisation like NMDC. A dynamic company was never stuck to same ideas, the same concepts and the same way of management. It has to adjust itself with the changing situation and growing expectations of the state and the community, else it would slowly sink into insignificance and die a natural death.

The directors agreed but were apprehensive because the task was enormous and challenges were huge. In fact, all of us were aware about the enormity of the task, but did not have full knowledge about the technological requirement of the job. We had to start from ground zero. We had to first learn and understand the technology, the economics, the logistics and above all the need to have innumerable clearances from various government agencies and support from the state and the community at every level. The job was challenging but rewarding, and strenuous but fulfilling. It amounted to trailing into an unknown domain associated with the anxieties of something unpredictable and pleasure of discovering something unknown. I, in particular, became a learner and I started learning from everybody I had acquaintance with. I tried to reduce the land requirement to the extent possible but found that getting an additional land of 800 acres was an inescapable requirement. We decided within a few days that our steel plant would be based on blast furnace route and we were soon convinced that there should only be one blast furnace with a size of more than 4,000 m³. We found that the railway rakes, which carried iron ore from Kirandul to Vizag, came empty from Vizag. So it was easy to import coking coal and use the available capacity to transport it to Nagarnar for feeding the coke oven plant. We started discussions with MECON and caught hold of every officer of MECON that we could do to learn about the technology of steel plant and the challenges involved. After acquiring some initial ideas, I went to Chhattisgarh to share our decisions with the state government. I was totally surprised to observe that our decision to set up a steel plant did not excite anybody. For all these years, NMDC had been blamed that it did not take any initiative to invest in value addition, but when we took the decision to set up a steel plant, the response was lukewarm. Everybody in the state indicated that we had already missed the bus. The state government already had agreements for getting two steel plants set up by private companies and a third steel plant by NMDC could be beyond the requirement of the state. However, by that time, I had made up my mind and decided to move ahead with the project. I had heard that there were some problems in respect of land acquisition by one of the private companies, which had not only identified the land and completed the process of land acquisition but also paid compensation to the concerned landowners, but they could not get the actual possession of land because of resistance from the local people and reportedly by the extremists. The area where they had decided to set up their plant was just about 18 km away from Nagarnar where NMDC was planning to set up the steel plant by acquiring another 800 acres of land beyond 1,000 acres that had been already under its possession. When I discussed our project with the senior state government officials at divisional and state level, some of them openly expressed their anxieties that the private sector steel project might not come up soon because of the local problems continuing in the area. They felt that a steel plant of NMDC was clearly a better option. They admitted that despite providing all support to the private companies to execute their project, progress was hardly visible. Some of them mustered up courage and informed the Chief Minister that the best thing for the state was to have a steel plant set up by NMDC which was committed to serve the social cause as well. One of them reported to have told the Chief Minister that NMDC had an established record of good and consistent performance in the state and that NMDC could be trusted much more than anybody else.

I could appreciate the apprehension and anxiety of the Chief Minister. He desperately needed a steel plant, a project which would provide employment as fast as possible to the people of his state who lived in the most backward region of India. He had already committed to the other companies that all support of the state would be provided to them. The steel plant project idea of NMDC had come too late. NMDC's steel plant would compete with the private sector steel plant which had shown determination to move fast. It would compete with the other plant for power, for water and for logistics. The Chief Minister was also not sure whether NMDC was really serious or was reacting to the loss of two rich iron ore deposits to private sector applicants. To me, the perception was different. I felt that having been engaged in mining in this area and having taken advantage of the huge resources, which had been thrown open to the company, it was the moral obligation for NMDC to set up a steel plant to add value, generate wealth and provide income in the region. This was not only a moral necessity, it was also needed for the company to sustain its business by fulfilling the aspirations of the people of the area. I found that NMDC, for many years, was not in a position to acquire a new mine. The scope of its horizontal expansion was not visible. It had to move forward in the direction of vertical integration through value addition. It would not only help NMDC protect its mines but also help it to become financially viable, socially acceptable and morally desirable. NMDC had to do so even if the state government did not encourage the move. I felt, in the case of NMDC, its own initiatives would bring everybody on board and ultimately, the community, society and the state would understand, recognise and appreciate the value of NMDC's effort.

With this conviction, I decided to move ahead. I also decided to meet the local Member of Parliament who was a tribal leader and highly respected in the area. When I met him, there was so much rain that the entire area was almost flooded. I got down from the car on the main road, crossed the waterlogged pathways and became almost fully drenched by the time I stepped into the veranda of his house. The MP received me with open arms. A glow of happiness was rolling out of his cheeks. He was excited to talk about our plan and said that he had been waiting for this day since long. I sought his support to go ahead. He was most vocal in welcoming the move. He said that it would bring about a change in the lives of people in the backward area of Bastar, which had needed such intervention from NMDC since long. He telephoned the Chief Minister of the state in front of me and told him that he was extremely happy with the initiative of NMDC. He said that if anybody did have the strength to set up a steel plant in Bastar, it was only NMDC and nobody else. He said that for any new initiative, rain was a favourable indication. It was good that we were discussing the plan while there was torrential rain outside as if everything of the past was being washed out and destiny was preparing us for a new beginning. Next day the local newspapers came out with the report in a big way. I felt that it was now time to directly go and meet the people to explain our plan and to share our dream for the future with them. I was very sure in my mind that the steel plant would come up only if the local people fully supported the initiative.

While I was thinking about it, a group of important people of the area, some journalists, a few panchayat chiefs and some local elders met me. They said that they would like to organise a public meeting for me to share our plan with the villagers. I could clearly feel that the wind had started changing in our favour. I could also see that expectations started building up around NMDC's steel plant.

Again there was heavy rain in the area when the public meeting was organised. Anticipating that, the organisers had covered a large part of the ground. But the number of people who gathered was so much that the crowd spilled beyond the covered area and took shelter under umbrellas. There was a building nearby with a long veranda, but that was fully occupied by a large police force, which without anybody asking for it, was posted by the district administration as a measure of caution, because any proposal to set up a new project involving acquisition of land could have met with resistance from local people. Such resistance were experienced in the area when another private company was trying to acquire land to set up a steel plant. The organisers had invited the Sarpanches of the nearby villages. I started the meeting by felicitating them. Each sarpanch welcomed me one by one. I then, started sharing with the audience our visions, plans, strategies and execution programmes. I told them that we had been a mining company for five decades. The company's relationship with the community had been excellent. Most of the employees working in the Bailadila project were from Bastar only. NMDC had been working in the area for decades in providing educational and health care facilities, creating infrastructural facilities by building roads and culverts, bridges and schools. Wherever possible, NMDC had stood by the people of the region, but then the basic problem of poverty had not disappeared. As an integral part of the society of Bastar, we felt that we would have to do something more if we had to do justice to our position in the region. I told them that we had made up our mind to set up a steel plant in Nagarnar where some land had already been acquired by NMDC in the past. I said that the steel plant had been conceived and planned keeping in view the need of the region. The steel plant would provide employment to about 10,000 people and along with indirect employment, it would bring about a change in the lives of at least 100,000 people around the plant. It would generate wealth and prosperity. It would inspire other industrialists to come forward and set up ancillary industries in the area. I explained that our initiatives on community work had provided relief to the people, but unless they acquired the capacity to earn in a productive way, the quality of their lives would not change. Our dream had been to create a situation when our initiatives on CSR would be redundant and people would not need them. There had been a number of large industries in the past in the backward areas of India, but many of them had not contributed towards changing the lives of people in the respective regions because while the plant was set up in the area, employees came from all over the country. Many such plants and townships glittered in the midst of darkness and poverty. Our dream for Nagarnar, was that every employee of the plant would ultimately be taken from the region. Most of them would be tribals. The employment will not come to them on account of compassion but purely on merit. We would build up institutions like an Industrial Training Institute (ITI), a polytechnic and an English-medium residential school to train the local people to acquire education and skill. Before the ITI, polytechnic and residential school are set up, not a single brick would be laid for the steel plant. I told them that some of the graduates, would be selected purely on merit from the tribal area and would be sponsored to the management institutes with full cost borne by the company in order to ensure that they become good MBAs. Our steel plant would take them as the first batch of management trainees. The polytechnics and

ITIs would cater to the needs of the skilled workforce. In addition, a large number of people will be trained in batches at the cost of the company, so that they could be readily engaged even at the construction stage by the contracting agencies. During the construction period, they would be exposed to further work experience, which would make them suitable for induction into NMDC for operating the plant when the construction work would be over. I shared with the crowd that this project would be executed only if they were willing to have the plant built in their area. NMDC needed another 800 acres of land and if the land was readily available, we would move ahead to set up the plant. In case, of any apprehension or doubt, I would clarify them. If there was any opposition, NMDC would not set up the steel plant at all. This steel plant was meant to be set up for the people of the region and for bringing about changes in their lives and also to change the future lives of their families and their future generations. The plant would also make steel with the resources already available in the area and the wealth generated thereby will go for the development of the state and the nation. The speech ended but not the rain. After finishing my speech, I came down from the stage and walked into the covered area. I had shared with them my ideas. Now I wanted to know their response. I went into the midst of crowd and got surrounded by them. The police personnel were apprehensive. They were not sure what was going to happen. I found that the faces around me started showing flickers of smile. Some of them made comments. After about half an hour, some elders came forward and said 'we want you to sit with us and watch some of our artists perform'. I could see some people ready with drums and sticks. We all sat together and spent another half an hour enjoying the programme. I was never confident about my fluency in Hindi, but I discovered that whatever communication I attempted had reached the minds of people. I decided to meet the Honorable Chief

Minister the next day on my way back to Hyderabad. Next morning, all the local newspapers made this meeting a front page news. When I reached Raipur and met the Chief Minister, I found him much more receptive. I later discovered that the gentleman was not only an excellent human being, he was a very mature politician and administrator. He knew how the jobs could be done without passing instructions. He smiled and said he knew about my meeting with the local Member of Parliament and also the public meeting held on the previous evening. He said that he was happy that ultimately NMDC decided to set up a steel plant. He said that it was necessary for the state, but NMDC had always been reluctant to come forward. He assured that all support would be provided by the state government. I requested him to grant an additional mine to NMDC, which would be captive for the steel plant. My reasoning was that, with the steel plant being set up, NMDC in national interest should not lose its position as a merchant miner. No part of NMDC's output, which was used by other steel-makers, should be diverted to NMDC's steel plant, because in that case, NMDC's steel plant will come up at the cost of other steel plants elsewhere in the country. In fact, even if NMDC produced much more than the current level of production, it was not possible to meet the demand for iron ore, which was growing in the country because of new steel plants coming up and old plants expanding their capacities. The Chief Minister understood the position and said that he would surely consider grant of an additional mine for the steel plant. Within a few months, it was done. Water and power supply was assured and notifications for land acquisition were issued. The state government came forward to support the project with all sincerity and with this the distrust, which had been built up between the state government and NMDC, evaporated.

With the initial hurdles being over, we jumped into action. A steel plant is not just an industry, it is a conglomeration of industries. MECON is a state-owned design and consultancy firm, which mainly caters to the needs of the Steel Sector. I personally requested the officials of MECON to work day and night for us. Some of our officers had camped in Ranchi to ensure that not a single day was wasted. Our first job was to issue a global expression of interest notifying to all concerned that our company was to set up an integrated steel plant of 3 MT capacity. There were nine packages in which the entire project was proposed to be divided. All packages represented turnkey jobs to be executed by one party or a consortium of parties. We invited response from the interested parties who would come to MECON and explain to a team of officers drawn from MECON and NMDC about the technology they would use, about their experience, about the type of terms and conditions they would insist on and a comparative analysis of different technologies and their opinions and experience about each one of them. Issue of notification for land acquisitions was about three months away and the environmental clearance and forest clearance were to be received several months later, but our job on selection of contractors for project implementation on turnkey basis had already started. It was a great session of education for our officers and engineers from MECON. For any project, the technology and the terms are specified by the owners. The suppliers respond to the tenders which are followed by discussions and negotiations on techno-commercial specifications, which take months, even years. I discerned that neither NMDC nor MECON knew the best of technology available in the world. We were not aware about the comparative advantages and disadvantages of each technology, equipment and their respective sizes, volumes, production capacities, etc. The difficulties or advantages, which might arise in respect of integration of one set of technology with others were also not known to us. I felt that we must know what were the equipment, machines and technologies available in the world and their comparative advantages. We must know the experiences in respect of use of different technology and different equipment in different steel plants of the world and then only would we be in a position to know and finalise our mind about what we need to do for our steel plant. We would also know who would be the suppliers of the right technology and equipment as per the specifications we frame up and shortlist them while issuing the final tender for inviting the price bids. We knew that for deliberations with the possible suppliers and turnkey contractors, NMDC did not have experienced people. We had picked up some of the officers from within the company with relevant background, but we also quickly hired a number of very experienced people, mostly retired from different steel plants. The entire batch of people so selected worked together with MECON for days and months thereafter to discuss with various parties coming over to Ranchi from different parts of the world and understanding from them the different options, which could be available to us. Another thing we did was that I sent a letter to the Central Vigilance Commission (CVC) informing them of our approach, which was totally new in Indian public sector industries. I explained that if we did tendering in the same way as others did, we would be doing so without knowing the options available with us, without learning about the technology available in the world, without capturing information about the latest development in the industry not only on the best production technique but also on fuel conservation, cost cutting, speed, automation, quality control and so on. We also explained that with these initiatives pursued parallely with land acquisition and environmental clearance we would save a lot of time and ultimately when the land was available and clearances were obtained, we would be ready with everything on execution side. There was no reply from CVC, but we knew that we were acting with full transparency. Only after a gap of about a year or so, we came to know of the outcome of this letter with a lot of pleasure and satisfaction. That will however, be discussed subsequently.

While work in Ranchi went on with full swing, the work in the field gained momentum. Our officers in the field were showing exemplary dedication. They worked on getting the notifications on land acquisition issued. They were carrying out a survey to find out the details of every family whose land was being acquired. They were planning ways to shift a public road, a temple and even a fully functional police station to get the land in one lot. They were working on identifying water sources and drawing water from those sources during construction and also for plant operation. They were working on rail connectivity and preparing for securing environment and forest clearance. However, most important work that they were engaged in was to ensure that, before any physical construction work started for the steel plant, a polytechnic, an ITI and a residential school were set up. Without any delay, the first two jobs stated above had started, but a problem came up with the third project. In order to set up a residential school right from the primary level for the tribal children of Bastar, we needed to ensure that the school had the best possible infrastructure, best facilities for recreation and of course, best teachers and teaching arrangements for the students. It was a difficult task because this had to be completed within a period of three months.

I decided to monitor the progress myself and started collecting daily reports from the concerned officers. Arrangements were made with the Dayanand Anglo-Vedic (DAV) Public School for running the school. The contractor was directed to complete the building in two and a half months' time including landscaping. Work started and people worked day and night. I myself had been the student of a residential school run by Ramakrishna Mission. I followed the same system of accommodation for students as we had in our residential school. Only difference was that the arrangements in the present school were much superior. There were four beds in each room with wooden floors, fans and cabinets. The hostel rooms looked excellent. I remembered that in our hostel room, we had photographs of Shri Ramakrishna Paramhansa, Swami Vivekananda and other monks. In this school also, I decided to have photographs of the great leaders of India in each room of the hostel. We put up photographs of Swami Vivekananda, Shri Ramakrishna Paramhansa, Pandit Jawaharlal Nehru, Subhash Chandra Bose, Dr Babasaheb Ambedkar, Birsa Munda, Mahatma Gandhi, Bhagat Singh and others. But the most difficult thing was yet to be done. We had to select the students. Many people came forward with suggestions. Even the district administration indicated that they would be happy to guide us on student selection. We thought it over and decided to do everything on our own. We had formed a few teams of volunteers, mostly belonging to the local tribal community. It included men as well as women employees. They were told to go to the villages in different parts of Bastar. Most of the villages were inaccessible. They had to leave their vehicles several kilometres away from the villages and walked up to the villages to speak to the village elders, the sarpanches and others. In each village, they explained to them our plan to set up a residential school. They told them about the steel plant that was being set up and explained NMDC's mission to man the plant almost exclusively by the people of the tribal region where the plant was being set up.

Despite the fact that I myself planned these initiatives, I was quite apprehensive in the beginning. Many of the villages were effectively controlled by the militants, who could warn the villagers not to send their children to NMDC's school. There was a risk that they could even oppose the construction of the steel plant and setting up of the proposed residential school, which would be an English-medium school with modern amenities. Nothing like that happened. There was no opposition from any villager. None of the villagers showed any apprehension about the proposal. From each village, boys between the ages of 6 to 10 were selected. There was no test. The only criterion was the willingness of the parents to send their children for studying in the residential school. Once the list was prepared, the parents were told that on a particular day and time, they should be waiting for a vehicle to pick them up from the main road near their respective areas and take them to the school. Our volunteers moved out in different directions in buses. They started picking up the students and their parents from the pick-up points. They did not miss any student because the parents and the students had reached the pick-up points much before the scheduled time. Most of the students came with their fathers. In some cases, some village elders accompanied them. They came from all directions from different districts of Bastar. The students came with eyes of surprise. None of them were properly dressed. Most of them were barefooted with uncombed hair. When they entered the school, the rooms were allotted to them. They were initially hesitant to sit on the beds, which were already kept ready for them. The parents sat on the floor and many of them started shedding tears. They said that they never had dreamt that their sons would be living in such a place. The teachers and volunteers assured and reassured them that their children would be looked after in the best possible manner. We had doctors to look after the students on a regular basis. We had already procured clothes, uniforms and even sleeping suits for the children. Immediately, after they took a bath and had lunch, the teachers took charge of them. The students had to be ready for school inauguration function, which was scheduled to be held only after two days.

Meanwhile, the notification for land acquisition was issued by the state administration. It involved acquisition of tribal land, which required consent of every land loser. The district administrator organised a public hearing before it could issue the final order. The outcome of the public hearing was quite interesting. When the meeting was over, I received a call from one of our colleagues in the site. He said that every villager present there, raised hand in support of land acquisition. However, after the meeting was over, about 150 villagers refused to leave the spot. They were insisting that their lands measuring about 150 acres was also to be acquired by NMDC because they wanted to be a part of the project. We needed some land for setting up a colony for the employees, which was not provided for earlier. I readily agreed and ultimately the additional land of 150 acres was also acquired. The next issue to be decided was payment of compensation to the land losers. This was a very complex issue. Though every state has some rates fixed and notified, those are taken in most of the cases as the lower limit of compensation. The situation in Chhattisgarh was not different though the rates had been increased substantially short while back. The practice was that the authorities at various levels would keep pressurising to pay more and the actual payment would go up as the level of discussion moved upward. Therefore, my thinking was that there should be negotiations only at the top level and not at various levels as is normally done. So we decided that whenever our officers are called for discussions, they would attend, but they would not agree to concede anything beyond the notified rate. The officers concerned were instructed that they must not leave the meeting until the meeting ended. They must be polite and show a positive attitude. They should only explain NMDC's point of view and express their helplessness to agree to anything beyond the notified rate. The first meeting ended in failure, but the second meeting was taken at the level of the minister, who was very annoyed with the officers concerned and ended the meeting at around midnight, telling that the Chairman of the company should speak to him. I spoke to him next morning. He quoted a rate, which was abnormally high, more than double the notified rate. He said that it was necessary to pay the land losers at that rate, as otherwise the extremists could provoke them into creating trouble. I suggested that I would like to discuss the matter with him in person. We decided a date for discussion. I knew that it was not possible to pay the rate the minister was insisting on and in any case, the matter would be referred to the highest level for final decision. I sent a request to the Honorable Chief Minister to take up the matter for discussion at his level. The Chief Minister agreed and the discussion was to be held on the same day on which I was to meet the minister. But, even before I went to attend the meeting convened by the Chief Minister, I went to minister's house and told him that I came to him as promised by me to discuss the issue. Since, now the matter had been taken over by the Chief Minister, it would be better if we go together to the Chief Minister and discuss the issue there. I felt that unless I did that, the minister could possibly feel that I was trying to ignore him, which was not my intention. The minister was quite supportive except that he possibly had some political compulsion to ask for an abnormally high rate of payment.

The Chief Minister was very pragmatic. He asked everybody what would be the sanctity of government notified rate, if they themselves violated it. He however, said that the company could surely be requested to pay higher rates if they could afford. He requested me to pay about 100 thousand rupees extra per acre to the land losers over and above the government notified rate apart from implementing in letter and spirit the relief and rehabilitation plan, which we had announced for the land losers and their families. I wanted to involve the Honorable Minister of Steel at this stage, because he actually represented the government, which was the owner of the company. The Chief Minister understood the logic. I flew to Delhi on the same evening and met the Honorable Minister next day morning. Then the Chief Minister's call came. The Minister of Steel took some time from Chief Minister and wanted to have my views. I explained to him that paying some extra amount to the villagers would mean nothing for a company like NMDC. Its impact on the project cost will not be very significant. He called back the Chief Minister and told him that the money was ultimately going to the farmers and therefore, he had no objection to pay a little more than the notified rate as suggested by the Chief Minister. The Chief Minister expressed his happiness and thanked him. Within 10–15 minutes, the matter was resolved. When I came out of the chamber of the minister and was walking through the corridors of the ministry, I got a call from Chief Minister himself. He sounded happy and relieved. He told me that he was personally thankful to me.

> When I look back, I feel there could not have been a better example of a win-win situation than this. The Chief Minister was happy because he could pay a reasonably higher amount to the land losers than the rate, which had been very recently notified. The Union Steel Minister was happy because it was he, who ultimately took the decision to grant better relief to the farmers. I was happy that the matter was so amicably settled. Our board was happy because already at the highest political level, a consensus was reached. The farmers were happy because they got what could be reasonably expected. Even the minister who asked for much higher amount was happy because ultimately he did not have to settle the issue and the final decision came from the top.

The payments were made during the next two days. The people were happy and there was no problem whatsoever from any quarter in immediately laying out a barbwire fencing. We now had to secure environmental clearance before going ahead with the construction.

Meanwhile, the school was to be inaugurated. Couple of days back, the children who would be studying in that residential school had been picked up from different parts of the region. The inauguration was to be done jointly by the Chief Minister of the state and the Minister of Steel, Government of India. The Chief Minister informed us that he would go on his own and wait to receive the Minister of Steel in the circuit house near the site. I accompanied the Steel Minister from Delhi. We had hired a small aircraft for going to the site directly from Delhi. It was to take about three hours to reach the nearest airport at Jagdalpur. I received the Honorable Minister at Delhi airport and escorted him to the vehicle, which was to take us to the aircraft through the tarmac. The minister boarded the car. I was to just get in. Suddenly an idea struck me. I told the Honorable Minister that he was going to undertake a three hour flight to a remote corner of the country just for inaugurating a school, which was operating at primary level for the time being. Did he feel that it was worth the trouble he was taking, especially being a Union Cabinet Minister which he was? The minister squeezed his eyebrows and asked me what I actually meant? I said, 'Sir, it is a small step of man, a great leap for mankind'. The minister was relieved and smiled. He only told me, 'come, get into the car.'

What I told the Honorable Minister was something, which was inspiring me always. In Bastar, education was the most powerful source of empowerment. Providing the best education in an English-medium school to 400 tribal children was nothing as compared to the vast magnitude of the problems, which Bastar was facing. But then, the steel plant, the technical schools, the residential school all were beginnings and the most important thing was that it was realised by every man in Bastar. They came forward to participate in these initiatives. The poor people, illiterate people, the people living far away from the so-called civilised world, the people who were surviving on rudimentary agriculture or hunting or by collecting wood and leaves of forest all knew that what was happening was good for them, their children and their future generations. They believed that everything was being done with due sincerity and they could trust those initiatives. They came in big numbers to attend the inaugural function. They did not have proper clothes to wear. They did not come as a part of a political organisation. They came as parents and relatives of the children who would be studying in the school. They came as members of the tribal community. They came to observe, to watch and to be sure that what was going to happen was good for them and good for the society. The children sang the inaugural song. They had come to the school just three days before the inauguration, but we all were surprised. Every child was looking different. We were surprised by watching the capability of the teachers and to see how fast they could motivate, inspire and teach the children to sing the inaugural song in a chorus. The song was the Hindi version of the famous song, 'we shall overcome, we shall overcome some day'. It was the most appropriate song with which the school could be inaugurated. Everybody got emotional.

The Honorable Chief Minister joined the chorus and told in his speech that he would do everything possible to ensure that the people of the region could overcome the problems of poverty and disease. He said that he would take these initiatives forward to every part of the state. Equally overwhelmed was the Honorable Minister of Steel, who also delivered an excellent speech, which touched the heart of everybody. After the inauguration, both the ministers entered a classroom. It was a modern classroom with all the facilities in it; decent furniture, large writing board, proper illumination, excellent ventilation and everything else what a modern classroom should have. Both the ministers ran together to occupy the first bench and they said that they had never had the opportunity to sit in such classrooms in their life. There was one more promise to the villagers, which was to be kept. We decided to have the same type of mobile medical van facility which we did have in the neighbouring areas of Bailadila mines. Two fully equipped medical vans were kept ready and they were flagged off by the Honorable Ministers. These vans were fully equipped with all types of facilities including facilities for carrying out some urgent surgeries. Doctors and nurses were available in each van and they were to go to all neighbouring villages on an appointed day and time to provide medical support to the villagers. The doctors used to keep complete medical record of every patient. Thus they knew the medical history of every patient they were attending. A central dispensary manned by senior doctors was also set up for providing treatment to those who needed prolonged treatments.

We got the environmental clearance very fast, but a small hitch withheld our work. There was a tiny patch of land within the area acquired by us, which was on record a forest land requiring forest clearance. The system had been that we could not act as per the environmental clearance, if a portion of land even if it was in one corner of the plot, happened to be recorded as forest land. So we had to wait for the forest clearance before the environmental clearance could be acted upon. I do not even now know what was the logic for such a system, but it delayed our project by more than six months. However, when we had received the forest clearance, people congratulated us, telling that we got the fastest forest clearance as compared with any other project in the country. We were told about certain horrific stories about projects that were stuck off for years for lack of forest clearance, almost in similar situations like that of ours.

Meanwhile, our team members working in association with MECON shortlisted a number of agencies and their associates for various turnkey packages. They immediately sent letters to the shortlisted parties advising them to indicate their price bids. Within next three weeks, the responses were received and it took less than a week after that to know who the lowest bidder was. For the first job to be awarded the final recommendation of our technical team was presented to the Board within a month's time. Some of the Members of the board were also representing the government in some large companies. They looked at the papers with disbelief. Their experiences had been that after the tenders were issued, there would be a series of responses and counter-responses, meetings and further meetings to finalise the technical specifications and commercial terms and then the price bid would be opened. In many cases, these exercises took more than a year and in some cases, the company had to go in for re-tender. They were surprised that within one month of environment and forest clearance, the company came up with a proposal to award a contract worth ₹10 billion through a limited tender as the parties had already been shortlisted through a process of effective deliberation. The Members of the Board did not find anything technically wrong because shortlisting of various parties was done only after issuance of a global tender. But then, the final quotes were received not from everybody but only from the preselected parties. Though some of the board members were initially hesitant, the proposal was finally cleared. A few of the members told me that since the procedure adopted was different from what others had followed, it would be better to take a legal opinion and keep it ready in the file so that if any questions arose later, the board members could take the protection of a legal opinion. It was however, not necessary. NMDC did not have to take the legal opinion. Within the next couple of days, I found a circular issued by CVC placed on my table. The circular used almost the same language, which we had used in our earlier letter sent to CVC in which we had told the Commission that in the age of fast changing technology, we had to recognise that we were not in possession of information regarding the latest technological innovations and best practices followed in steel industry across the globe.

Therefore, our intention had been to first issue a global Expression of Interest primarily in order to learn from various respondents about the latest technologies available in the world in the field of steel making and then only to finalise our technical specifications and commercial terms or shortlist the parties satisfying those requirements. The price bids were to be invited only from the shortlisted agencies for doing the final selection. For months together, we had not received any response from the CVC, but the circular, which came from CVC, had something for us to rejoice. CVC in very categorical terms advised all the public sector companies to follow the procedures that we adopted. It reproduced the language of our letter almost word for word to state that in a world of ever-changing technology, it was important that the best technologies and practices in the world are captured before placing the orders. This gave us a lot of satisfaction. We were reassured once again.

> If we work with transparency, objectivity and clear logic and take decisions on the basis of the same, we need not necessarily follow the rules, rules will be framed on such decisions, because rules are nothing but codified experience.

As the construction work progressed, the sleepy suburb of a tiny district town called Jagdalpur suddenly attracted global attention and drew people from across the world. Contractors started coming from Germany, Italy, China, USA and from almost every part of India. We estimated that at the peak of construction, at least 10,000–15,000 people would be working together in 1800 acres of land to build a modern integrated steel plant, which would add 3 million tonnes of steel to India's steel production. We were confident that the steel plant would not only be an example of fastest possible land acquisition, it would also be an example of fastest possible execution of a project involving an estimated expenditure of ₹155.25 billion. We knew that the biggest problems, which the contractors would face, would be to secure skilled and semi-skilled manpower from the local area. Each contractor would be required to bring in people from elsewhere and provide them with accommodation and other facilities. We had earlier decided to adapt a new strategy to address this issue. We picked up local youth and started sending them to Hyderabad for skill development in batches. The entire cost of training was borne by us. The training module was so prepared that after a few months when they finish their training, they would have the minimum skill to work as construction workers in the project. This gave the contractors a big relief. Some students from ITI, which we had set up, had started passing out. Boys were available from other ITIs as well. Some of them were also funded by NMDC. A large number of skilled and semi-skilled manpower was thus available to the contractors when they came to execute the work. They did not have to get workers from other areas; they did not have to spend on their accommodation; they did not have to bother about the seasonal absence of those workers. They got trained people in the locality itself. We only told the contractors that they must engage them intensively, so that the workers could acquire valuable experience. Being associated with the plant from the construction stage itself, the workers would secure invaluable knowledge and experience about the concerned equipment and could be usefully engaged by NMDC later, when the operation commenced. It would be a win-win situation for everybody. Another advantage of the scheme was that after land acquisition, the land losers were not required to wait to be productively engaged till the plant was operational. They got an opportunity to work in the plant right from the construction stage and this had its own positive effect in maintaining the morale of the people as also social harmony.

In the capital goods sector, for any new plant, the biggest problem is that the EBITDA margin should be large enough to take care of interest on capital employed and depreciation. In an old plant on actual or notional basis interest on capital employed would be very low and depreciation would be almost non-existent. Therefore, as compared to a new plant, the old plants will always have a better financial viability. We had this benefit in our mines, which were constructed long back with low investment. However, this advantage would not be available for the new steel plant where the investment was more than ₹50 billion per million tonne of production pa and the minimum capital cost loaded on production per tonne of steel was ₹5,000. We had also decided not to take advantage of our in-house iron ore in our cost calculation and therefore our transfer price of iron ore from the mines was equal to the market price. We felt that the steel plant on stand-alone basis should be a profitable venture. We had the best of technology available, which would give us benefit in the manufacturing process, both in terms of cost and quality, but that was not adequate to generate the EBITDA margin that we desired to achieve to make the steel plant profitable from the very beginning. For that we needed to cut down the cost, especially the cost of raw materials. I had an idea in mind, which I decided to implement.

We had a large quantity of tailings dumped in our tailing ponds in our Bailadila mines. In fact, the tailing ponds were so full that it was becoming increasingly difficult to dump further tailings there. We felt that if we could retrieve some of the tailings from the tailing ponds and use it for pellet making in the steel plant site then the pellets would not only substitute high-quality lumps of our mines, which had huge demand in market, these would also be low in cost because the tailings they would be manufactured from would have hardly any commercial value. This would cut down the cost of raw materials used for steel-making substantially. We calculated that the composite cost of iron ore feed to our steel plant would be quite low as compared with other steel plants not only because we would be using low-cost pellets but also because of the proximity of our steel plant to our mines on account of which we would be saving huge cost on logistics.

We found that our Steel Plant would give an internal rate of return (IRR) of 19 per cent, which was very high as compared with other plants in India. In fact, this was based on the current price of steel when the market was under recession and it could go up further. We only needed to strengthen the logistics from our mines to the steel plant. We requested the railways to double the railway line from Bailadila to Jagdalpur, which was on plain land. This did not involve any engineering challenges, which were there in the remaining part of Kothavalasa–Kirandul line which had 58 tunnels and 84 bridges. The railways replied that they could do so, if NMDC funded the project. We agreed, because doubling of Kirandul–Jagdalpur sector of Kothavalasa–Kirandul line was necessary not only for the Steel Plant but also for evacuation of iron ore from Bailadila. We decided to sign an MoU with the railways giving commitment for investment from our side and the Railway Board agreeing to give us benefits in terms of freight concession over a period of time.

Another important decision, which we took, was to lay a slurry pipeline for pumping iron ore slurry including tailings from Bailadila to the steel plant site near Jagdalpur for manufacturing pellets over there. The idea was to take out tailings from the tailing dam, as explained earlier, blend it with iron ore fines, beneficiate it at Bailadila itself and pump beneficiated iron ore through the pipeline to the Steel Plant site for manufacturing pellets and using it in the steel plant. The pipeline was to give us a huge relief on two accounts, reliability of transportation and substantial reduction in cost on logistics. Consequently, the estimated landed cost of iron ore to be used in the steel plant went down further making the entire project more profitable.

We wanted to build a plant which would not only use the best technology and show best performance but would also be excellent in looks with good landscape, extensive plantation and well-designed layout. Therefore, even before the construction work started, we got the entire area levelled and started doing plantation by creating a green belt within the plant area. Some of the major construction work started almost simultaneously. By the time I superannuated, construction on most of the packages were in full swing.

After about six months of my retirement, one day I got a call from the Chief Minister's office informing me that the Chief Minister wanted me to come over to his place and meet him. I complied with his request immediately. He told me that a few days back he was flying over the plant area in a chopper. He was extremely excited to see how a huge plant had started emerging from nowhere in a remote part of the state. He told the pilot to have a second round over the plant area. He said that it was a thrilling and satisfying experience for him to see the mammoth structures coming up, chimneys getting erected, dumpers and cranes moving all around. He said that while watching all these, he was remembering me. He was thinking that my determination to set up a Steel Plant in Chhattisgarh had finally come true. He wanted to share his feelings with me and that was the cause of this invitation. I had no words to express my feeling. I felt that it was the best reward I could ever think of. I immediately remembered the Member of Parliament from Jagdalpur who had given us the required support in the initial days for getting the Steel Plant set up. He was no longer alive. I suggested that the new road constructed by NMDC semi-circling the steel plant be named after him. That was the minimum we could do for a great man like him. I knew however that the people of the area would in any case always remember him as their own man who had the right vision and right initiative.

In my long years in public sector, I have come across many politicians at various levels. Some were ministers, some were People's Representatives, some were Trade Union Leaders and some were Social Workers. I have seen that all of them were invariably happy when something good happened for the people or for the society. I found that when they came into politics, they were surely inspired by a feeling of doing something good for the country and for the people. Long years in politics and the struggles to remain in power might have compelled some of them to make some compromises here and there, but I have not come across any politician, whose eyes did not light up when any good thing happened for people. Another thing that I observed in politicians is that most of them have very strong common sense and excellent grasping power. Even in less educated politicians, I did not observe dearth of common sense and power to understand.

4

WHEN MARKETING BECAME THE KEY

While we were planning for growth and stabilisation of the company through diversification and forward integration, the global economy was passing through a phase of volatility. The impact was obvious on our company as well from the last quarter of 2007 when I had just joined. The steel industry in India as well as in China and Japan was witnessing a huge boom. Not only did China expand its capacity of making steel, its existing steel mills enlarged their production to capture the growing demands in the domestic and international market. In India, steel prices moved up to such an extent that steel, which contributed only a small weightage in the Indian consumer price index, had to be removed from the commodity basket to prevent its disproportionate impact on the index. Consequent to these developments, the price of iron ore increased substantially. China became almost the single destination for seaborne iron ore traffic across the world. It impacted the iron ore markets in Australia, Brazil and also India. From India itself, about 100 MT of iron ore was shipped to China during that year at spot prices, which were much higher than the long-term prices at which the contracts of NMDC were signed with Japanese steel mills at the beginning of the year. In the domestic market, the long-term customers from NMDC got a huge benefit due to low iron ore prices, which were based on the long-term contract prices with the Japanese but were substantially lower than the spot prices of ore in the international and domestic markets. The situation had thrown a challenge to NMDC. The question was whether NMDC should retain the long-term price of iron ore for its domestic customers until the end of the financial year or whether it should increase the price of iron ore through a mid-term revision of the price for which there was a provision in the contract, though it had never been invoked.

I had just joined NMDC, but my mind was very clear on the fact that we should go by the contract and increase the prices. It was not only legally but also morally correct, because the long-term customers were already making enough money, which they could share with the supplier. Since the concerned clause of the contract was being put into use for the first time, we quickly took a legal opinion and became sure that our actions were in the right direction. With effect from 1 December 2007, the price of iron ore was increased by 47 per cent. There were petitions and representations from the customers who decried the move. Some said the action was illegal; some said that the concerned clause in the agreement that NMDC could do mid-term revision of prices was by itself unlawful and should be immediately repealed. Some said that the act was immoral because there was no increase in prices for long-term customers in the export market, while only the domestic customers were penalised. The fact was that in export contracts, there was no provision for mid-term revision of prices.

This clause existed only in domestic contracts. However, there was no pressure, no suggestion and not even an indication from the ministry or minister regarding the price rise effected by NMDC. I could clearly realise what autonomy and professional independence meant and relished it.

The price of iron ore increased in the domestic market, but the boom in steel industry continued. Next year, the long-term price of iron ore itself increased substantially. Price rise of its product brought a huge profit for NMDC and gave it confidence about its value and significance. Things looked perfectly bright as we stepped into the new financial year.

However, far away from us, a cloud of severe recession was looming large on the global economy. Nobody expected that the degree of crisis would be so huge and that its impact would be so widespread. The US economy suddenly sunk into severe recession. The fall of Lehman Brothers symbolised the collapse of the financial market in the USA, which spread shivering waves of uncertainty across the globe. The crisis originated from the overheated sector of housing finance. With sudden fall in the prices of dwelling units in the USA, the banks which had advanced huge sums of money against mortgage of those assets suddenly sunk into crisis, and its impact was felt by every institution which had bought those mortgages from the lending banks. The housing market disaster impacted the business of construction. The banking sector crisis restricted capital inflow in the economy and, as a consequence, the steel market crushed heavily. In India, the steel prices fell by more than 20 per cent in one go. The whole situation in the iron ore market became just the reverse of what it was in the previous year. The steel sector was losing money and iron ore prices became higher under long-term contracts than what the steel industry would afford. The situation in China was not that bad because of a huge domestic market. China did not allow the steel mills to cut down the production; it only decided to pre-pone some of the projects which were decided to be implemented much later. Some bridges, tunnels and buildings which were to be constructed after 2011–12 were taken up for construction immediately. The actions of the Chinese government helped the steel industry in China to overcome the crisis to a large extent.

In India, however, the crisis was not easing out. It was, rather, getting aggravated. I sat with some of the senior officers to take their views and was appalled on seeing the helplessness in them. I told them that the value of the management was tested in crisis. We would fail in management if we surrender to the situation. We decided to take certain actions immediately. The first was to rationalise the price. The clause in the long-term agreement with the domestic customers regarding mid-term revision of the prices existed in any case, and this was time when we needed to do corrections in prices to enable the customers to lift material from us. Iron ore is a bulky material and occupies a large storage space. Iron ore mines are, therefore, operated on the basis of evacuation at the same rate of production providing for space for stocking not more than a week's production. Low lifting of iron ore by the customers, therefore, had a direct impact on NMDC's production. In November 2008, the sale of iron ore was just 1/3rd of the normal sale. As a result, stock yards were full and production was low. In some areas, the production had to be suspended completely. While deciding that we should reduce the price of iron ore, a question came, by how much? We had a formula to decide the percentage of reduction and we decided to use it. We also took note of the fact that the price elasticity of iron ore was low and, therefore, unless the reduction in price was substantial, we would lose only revenue but would not be in a position to encourage the customers to start lifting the material. We wanted to increase sales to a normal level and for that a visible cut in price was essential.

We called a special board meeting and all the board members were convinced with the proposal to cut down the price of iron ore by 25 per cent. This was immediately given effect. An interesting incident comes to my mind in this connection. After we increased the price in December 2007, some customers met me and requested that the price increase, which was effected for domestic customers, be withdrawn. They had also sent written representations to us with legal and moral arguments to repeal the clause in the agreement, which was invoked by us while increasing the price during the previous year. Now when the market started reeling under recession, some of the same group of customers met me and insisted for reduction in price by way of mid-term revision from the rates fixed as per the long-term contract. Putting on my best poker face, I told the customers that on the basis of their previous representations NMDC had already decided to drop the concerned clause from our contract for doing any mid-term revision of price. They were visibly stunned. They had no answer. I watched their gloomy faces but remained silent. Immediately after the board approval for price cut, each one of them was informed of the decision which brought unexpected relief to them. Many a times, our response to situations come from a reaction. Nobody expected a year back that there could be recession in the market and, therefore, pressure came to remove a clause from the agreement. This ultimately came to help the customers, the industry and, of course, NMDC during the next year.

We took a series of other steps. Our contract with the customers provided that a certain percentage of the agreed quantity would have to be lifted by the customers if they wanted to retain their long-term status. We decided to send letters to every long-term customer and reminded them of their obligation. We cautioned them that if they failed to lift the material as per the agreement, their long-term customer status would be in jeopardy. This worked well because long-term customer status gave raw material security to the users of iron ore, which was always important for the process industry. Long-term customers also received a substantial benefit in prices because the long-term prices were generally lower than the prices in the spot market, and that was the reason for which due to excessive volatility in spot market prices, adjustment of long-term prices became necessary. Our letters to the customers were not only a discreet threat but also reminded them of the benefit, associated with their long-term status.

In addition, we also decided to issue a press advertisement for enlistment of new customers. Our policy was to allow some of the customers to lift material on spot prices on regular basis and, subject to their performance, they were assured to be given long-term status from the following year. The advertisement had two impacts. First, the long-term customers got a message that if they did not lift material, they might lose long-term status because new customers were already in queue to fill up the void created by their exclusion. Second, to encourage potential customers to come forward and pick up material with a hope that they would ultimately get long-term status, a benefit which was much sought after in the steel industry. The response was overwhelming.

Meanwhile, in October 2008, when the meltdown of global economy had just started, I decided to visit China along with the officials from the Ministry of Commerce and MMTC Ltd, which was the canalising body for export of high-grade iron ore. We spent about a week in China visiting a number of integrated steel plants. NMDC's trade relationship had historically been with Japan and South Korea with whom it used to have long-term agreement for supply of iron ore. Japanese steel mills had been associated with NMDC right from the stage of discovery of iron ore deposits in Bailadila. They had assisted India in building the KK line for transportation of iron ore to Vizag and in constructing the outer harbour in Vizag port for easy transportation of iron ore from Bailadila to Japan. All tie-ups for exporting iron ore to Japan were on long-term contracts. The same system existed for export to POSCO in South Korea. Normally, NMDC's material was not exported to China, which used to procure iron ore from various countries including India at spot price. Sensing that the global crisis may affect Indian steel mills and its iron ore miners, I had decided to sign a few non-binding MoUs with Chinese steel mills. This move was welcomed by the Chinese steel mills, which had been hoping to have some sort of tie-up with NMDC for quite some time because of the high quality of ore that the mines of NMDC produced. While signing the MoUs, my plan was to use the MoUs only if we failed to sell the material in the Indian domestic market. These MoUs turned out to be very useful within the next couple of months. In November 2008, when NMDC was able to sell not more than 30 per cent of its production which led to the mines getting choked, we decided to export iron ore to China in terms of the MoUs we had signed and ultimately, till March 2009, we exported about 5 MT of iron ore to China.

In addition to the four strategies mentioned earlier, we decided to expand our stockyards and build new stockyards for stocking the material produced from our mines, so that our production was not affected due to any hold-up in delivery for a few weeks or more. This five point strategy reaped dramatic results. In December itself, the sale of iron ore became three times than that of November. NMDC ended the year with record profits.

Incidentally, this phase of recession gave us a new insight about the workings of Japanese, South Korean and Chinese steel mills. When the price dropped drastically and the spot price became much lower than the long-term prices in the global market, we had a discussion with the Japanese steel mills. During the discussion, the Japanese never talked about price at all. They only spoke about phasing of supply to some extent to facilitate the steel mills to plan for rationalising the level of production. They did not utter any word on price as they knew that long-term prices had to be honoured. In fact, there was no failure on their part to pay at long-term rates even during the worst phase of recession. The South Koreans talked about the difficulties and hardships that they were facing, but knowing that Japanese never talked about price, they refrained from discussing the price issue. The Chinese steel mills were supplied material on spot prices, but when the price fell further, we suddenly started getting complaints about quality and grade, which we had to settle through umpires.

When the global crisis impacted India the most, the steel industry was possibly the worst hit. In 2007, the industry was experiencing a boom. Nobody expected that within the next one year the boom would be followed by a recession, which would not only affect India but other countries as well. Two global organisations decided to hold a seminar in Hyderabad to have a daylong discussion on the issue. This seminar was attended by the industry stalwarts, top bureaucrats and others. I was given the task of introducing the subject. I remember having told two things in the seminar. First, that the recession in India would not last long, and by the middle of the following year, the situation would start changing. Interestingly, this forecast came true, though the steel industry did not go back to its original level of profitability even during the next few years. My second observation was that the commercial system built by NMDC had withstood the test of time. The strategies which we had adopted to combat recession had all been based on the systems and procedures that already existed in the company, and the credit for that should go to my predecessors who had taken great pains to build a structure of sales and marketing that could survive even the most volatile phase of the global iron ore market. The subsequent speakers agreed and some of them observed that the entire crisis had been created due to the extremely short-sighted approach of US banks in particular and the global financial sector in general, which believed that the housing sector prosperity in the USA would perpetuate. On the other hand, NMDC's sound commercial systems provided for all possible eventualities and that's why the company could withstand the impact of recession on its sales and production.

When I analyse the situation in retrospect, I find that our success in overcoming the problem of recession cannot fully be attributed to the commercial policy of the company. It was surely useful and important, but equally important was possibly our timely response to the situation and our drawing up a five point plan as soon as the crisis had set in. The right plan and its timely and meticulous implementation helped us not only in coming out of the crisis but also in posting record profits in 2008–09, that is, during the year of recession. It was evident that the management's response to a situation of crisis is an essential requirement for overcoming the crisis: case in point being NMDC, where the response of the management was prompt and appropriate and the company came out even stronger.

In 2010–11, two important developments took place. The first was the change in the pricing structure in international trade of iron ore and the second was the imposition of export duty and differential railway freight for the export of iron ore from India. The first change was inevitable. For quite some time past, the long-term price of iron ore was being determined by the Japanese steel mills and POSCO in South Korea. Due to the huge growth in the steel sector in China, China became the major importer of iron ore, and it started importing iron ore on the basis of spot prices, which varied on day-to-day basis. As the demand grew, the spot prices started indicating an upward trend almost consistently. Fall in prices in spot market happened only in 2008–09, which was the year of recession. Otherwise, the long-term prices, which were determined at the beginning of the year were almost always lower than the spot prices. The iron ore producers and exporters, therefore, found it inconvenient to stick to the long-term prices. They insisted that long-term contracts be based on quarterly price adjustments, the price of each quarter being fixed on the average quarterly price at spot market. The result was that even in long-term contracts, the prices increased substantially, and this came as a boon to NMDC because both in domestic and export market, the price at which NMDC was selling iron ore became about two times the price charged previously. The company started experiencing a new height of affluence and it gave us the courage to get into new ventures and initiatives.

When the prices of iron ore increased substantially, the profits of the miners went up rapidly. Various authorities thought of extracting a part of the high profit earned by the miners. The finance ministry felt that a part of export realisation could be taken out as export duty and that would help in mopping up a good amount of revenue. The rate of duty, which was initially 5 per cent, ultimately moved up to 30 per cent, making much of the export business unviable. The railways felt that they, as a carrier of iron ore to the ports, had a right to extract higher revenue from iron ore exports. The rate charged for export of iron ore was decided to be made more than 2½ times the rates charged for the domestic transportation of iron ore. In fact, the railway freight from mines to port became the highest freight to be charged by any railways in the world. It constituted almost 35 per cent of the free on board (FOB) value of exports. As a result, roughly out of the total value realised on FOB basis for export on iron ore, 30 per cent was to go towards export duty, 35 per cent towards railway freight, 10 per cent towards royalty and another 5 per cent towards handling charges. The miners were left with just about 20 per cent of the FOB value of its exports. In the process, exports became un-remunerative. Export of iron ore, which crossed 100 MT in 2009–10, came down to just about five MT four years later. But this did not help the steel industry in India, which was lobbying hard to stop the export of iron ore from India. Due to intervention of the Supreme Court, a number of mines were shut down on the grounds of either illegal mining or environmental violations. Therefore, Indian domestic iron ore production declined by more than 75 million leading to an increase in the domestic price of iron ore. Meanwhile, in respect of the pricing of iron ore, the industry lost its benchmark. Due to insignificant export, the global export price of iron ore could no longer be accepted as the benchmark for the determination of the domestic price, as was the case earlier. The iron ore miners started adjusting their prices on the basis of domestic supply and demand, which was not necessarily favourable for the steel mills. As a consequence, India, which was a net exporter of iron ore, now became a net importer as the coastal steel plants in particular found it preferable to import iron ore.

I personally feel that this whole episode was yet another example of how wrong policies and wrong decisions affect the interest of the nation and its economy. When the iron ore prices were high, if India could have continued to export iron ore, its foreign exchange earnings on account of export of iron ore could have been as high as $10 billion pa. India lost that opportunity. Everybody in the industry knew that iron ore prices were set to fall in the near future. India was always free to import iron ore when the prices were low to meet the needs of its steel industry. Instead, India decided to impose duties and lost a valuable opportunity. The Railways were not justified in imposing differential freight on the export of iron ore. In fact, it all started with some sort of congestion surcharge during the busy season and was later converted to be an instrument for earning revenue at the cost of industry. In the process, neither the government got its revenue from export duty, because there was hardly any export, nor the railways earned higher revenue from export of iron ore because only the Goa mines, which did not use the railways, could do whatever exports were being done from India.

While all these changes were happening in the business environment in India and abroad, two interesting developments took place in NMDC. Suddenly, the Karnataka Lok Ayukta came out with a report. It said that by exporting iron ore to Japan and South Korea on long-term prices, NDMC had lost money by giving undue benefit to those long-term customers. It went on to say that since NMDC was selling iron ore at a price lower than the prevailing spot price, there were reasons to believe that the company was parking illegal money elsewhere. In the three pages of its report, the Lok Ayukta raised very serious allegations against the company and its report was circulated to all concerned. The next day, *The Times of India* came out with front page news on Lok Ayukta's report against NMDC. Our pride in being an honest and transparent company got a rude shock. I called a press conference in Hyderabad on the same day and explained how the long-term prices had been different from the spot prices and how the global long-term trade was always based on the long-term pricing structure and not on spot prices. Pricing is always a difficult and monotonous subject. Therefore, I had to explain the background, the mechanism and, ultimately, the system globally followed in seaborne trade of iron ore. Next day, a press conference was called in Delhi. There was huge attendance. I told the media, at the outset, that I would not leave the hall before the last question was answered and the last doubt dispelled. I felt that the report had thrown a challenge to NMDC and put our reputation under a cloud. It was our duty to explain the truth and to convince everybody that what was mentioned in the report was practically not based on facts but on some simplistic assumptions, which a reputed body like Lok Ayukta should have avoided.

The press conference continued for hours followed by interviews to TV channels. Almost all the newspapers carried bigger headlines the following day covering the press conference and supporting the views expressed by NMDC. After about two months, the Supreme Court decided to suspend the mining operations in Karnataka until the issues relating to illegal mining were resolved. Allegations of mining in violation of environmental norms were thoroughly scrutinised for each mine, and a report with regard to the same was submitted. The Supreme Court decided to continue suspension of mining operations in Karnataka. It, however, decided that only NMDC's mines in Karnataka would be allowed to continue its operation with the provision that NMDC would ramp up its production. The Supreme Court set a target that NMDC mines in Donimalai should produce 12 MT pa instead of the 7 MT earlier, so that the needs of the local steel plants could be satisfied. While the Supreme Court was pronouncing its judgment, one social activist who had filed a PIL on illegal mining suggested that NMDC itself was not an Honorable company because there were severe allegations made by Lok Ayukta against NMDC's management. The Court rejected the objections made by the concerned person and a reference was made to my press conference and TV interviews in which I had clearly explained the distinction between the long-term price and spot price and that the Lok Ayukta had clearly erred in not appreciating the difference. It was stated that not only was my logic sound but my body language had also been clearly convincing. The court asked our lawyer to get a confirmation from the management whether we could produce 12 MT pa from our Karnataka mines. The lawyer hurriedly came out of the courtroom and telephoned me. I advised him to confirm that we would produce one MT of ore per month, that is, 12 MT pa, to meet the needs of the steel plants in Karnataka. Immediately after the judgement was passed, we took all measures to ensure that the production picked up from the following day itself and our directors physically led the drive from the front.

Meanwhile, the impact of high export duty and high railway freight made NMDC's net realisation from export so low that it became unviable to export iron ore to Japan and South Korea, despite the fact that the long-term prices, which were now decided on the basis of average spot prices on quarterly basis, were quite high. The export contract with Japan and South Korea was due for renewal. I informed the government that it was not possible for NMDC to export and realise much lower than what it could get by selling the material in the domestic market. I did not receive a formal response to my note. There were, however, oral communications to me that, in national interest, we should export to Japan in particular. My point was very simple. If it was in national interest, there was no problem for NMDC to do anything that it was required to do, but as a commercial organisation it could not do anything which was directly against the commercial interest of the company. Therefore, if at all we had to export, we needed some written communication from any responsible authority with which we could justify our actions. No such communication came and our exports to Japan were kept in abeyance until I superannuated from the company.

The period of 2007–11 was a period of turbulence in the global economy. It was a challenging time for those in the steel and mining sectors in particular. For the first time in the history of NMDC, the company was exposed to those challenges. Our determination to fight against all odds and confront the waves of new situations coming up in the domestic and overseas markets gave us the maturity, which contributed towards making NMDC a mineral giant with strong fundamentals. I find myself incredibly lucky that I got the opportunity to lead the company at this phase. The company was not only handling the external challenges, it was also evolving itself from being a closely held company to a public company having equities held in the public domain. Its disinvestment was planned and executed in 2009–10, and we sailed through the process of disinvestment with a great success.

If we look into history, we will find that a boom is rarely followed by recession in such a quick succession as it happened in 2007–08 and 2008–09. A number of banks and financial institutions collapsed in the USA because the financial sector was the worst hit. But the impact on other industries, especially steel and infrastructure, was also enormous. In fact, the steel industry during the next five years could not retrieve the position that it had lost in 2008–09. In case of iron ore, the situation in India became increasingly difficult for environmental and other reasons, and a large number of mines had to be shut down. Indian production of iron ore fell by more than 35 per cent. Indian export of iron ore fell by more than 90 per cent. But NMDC not only survived, it prospered. The company which in 2007 had the net worth of ₹30 billion, reached a net worth of about ₹250 billion by the end of 2011–12. NMDC converted every threat into an opportunity and every weakness into a strength, which was very evident during the public issue of its equity in 2009–10, wherein the Government of India mopped up a massive amount of ₹99.30 billion by divesting just 8.38 per cent of the equity of the company with a face value of just about ₹330 million.

The entire process of NMDC's equity sale was handled by the company in association with the administrative ministry and the department of disinvestment. The value of equity depended on the worth of the company and the confidence of the investors on its management. Roadshows were conducted in different cities of India as also in Hong Kong, Sydney, Singapore, New York, Boston, London, Frankfurt, California and a few other places. A mining company operating in a remote corner of India was by this time a known name, not only in India but also abroad. Disinvestment brought it into the global limelight. Every investor was thrilled on going through the balance sheet of NMDC. They were excited to see its net worth. The only point bothering them was high offer price for NMDC's scripts. The offer was ₹300 for equity of ₹1. Many investors, especially those from overseas countries, had questions on NMDC's diversification into steel, as steel would be less profitable than mining. They had to be convinced that the steel plant being set up by NMDC would ultimately help NMDC not only to diversity but also to strengthen its position as a mining company. We ran through a very hectic schedule and for hours together presented the company and its strengths, and opportunities to the investors. We clarified their doubts and answered their questions. The money collected through disinvestment went to the government, but our effort in telling the world about the value of the company we represented did not go in vain. Along with the largest mining companies of the world, NMDC came to the centre stage of global mining and became a known entity in the mining sector with offers of acquisitions, joint ventures or participation in many global mining initiatives flowing to NMDC from all corners of the world.

When I look back, I feel that disinvestment through public issue was a huge improvement from the strategic sale adopted by the Government of India till 2004. In a strategic sale, the government loses the management control of the company. Strategic sale is in reality privatisation with the government being reduced to minority partnership or no partnership in the equity stake of the company. Since only a few parties could participate in a strategic sale, because of the size of the investment and other factors, the competition was very limited and, therefore, the value realised was much lower than what the assets could have otherwise fetched. This became evident when the government adopted the stake sale approach and started divesting only a small percentage of equity in the market in a single lot and, thus, collected a huge sum of money. With this, the government not only retained management control but was also able to mop up a larger amount of money than what strategic sale could have fetched. The success of government stake sale of NMDC is an example of that.

5

STRUGGLE TO EXPAND

As a mining company, NMDC's strength was always undisputed. The only issue that the company faced was that of stagnation. Geographically, it was confined practically in two states, namely Chhattisgarh and Karnataka. Its existence in other states including Himachal Pradesh, where the company had a concession for limestone mining, Jammu and Kashmir, where it had a mining lease for magnesite mining, or even Madhya Pradesh, where the company had an operating diamond mine, was insignificant. Product-wise, the company was practically a single-product company involved in the mining of iron ore. The contribution of diamond in the total turnover of the company was insignificant. Volume-wise, the company was in the production range of 20–30 MT of iron ore because the difficulties in logistics had restrained the company from increasing its production, especially in the Bailadila sector. A strong company like NMDC needed to expand its activities. It needed to get new mines, increase in production and get into backward and forward integration; it had to diversify and grow vertically and horizontally. With the steel plant coming up, the company succeeded in achieving its objective of forward integration.

Simultaneously, it was decided that the company should go into beneficiation and pelletisation in order to add value to its products and, ultimately, to achieve one of its core objectives of mineral preservation. The first pellet plant of two MT capacity was planned to be set up on the same premises where the steel plant was to come up. Setting up of a pellet plant along with the steel plant was a strategic decision. While setting up the steel plant, the company's conscious decision had been that the steel plant should come up to add value on a stand-alone basis, and the gains made out of steel plant should not amount to transfer of value from mining to steel-making.

The decision, therefore, was that hardly any material would be diverted to the steel plant as raw material from the existing production. A part of the required raw material would come from the new mine, which was recommended to be allotted as a captive mine for NMDC's steel plant but was to be developed as a joint venture project of NMDC and the State Mineral Development Corporation. The steel plant was to draw material from the new mining project purely at market price. About 30–40 per cent of the requirement of raw material would be met from the Bailadila mines, from the wastes that the mine generated. Only a small quantity of fines was to be blended with the tailings for producing good quality iron ore concentrates to be pumped to the pellet plant at steel plant site. The pellet plant would not only feed the blast furnace but would also sell a part of its products in the open market to meet the needs of other sponge iron plants in Chhattisgarh, which required more raw material than NMDC's Bailadila mine could supply. The second beneficiation-cum-pelletisation plant was decided to be set up in Karnataka near NMDC's mines in Donimalai area in Bellary. This plant was to have a capacity of 1.2 MT pa. The main purpose of setting up the pellet plant was to use the tailings in Donimalai where the tailing ponds were already full, with the total accumulations exceeding 20 MT, and evacuation of materials from the ponds had become necessary from an environmental point of view. The tailings as was planned in Bailadila were to be blended with good quality fines, beneficiated and then pumped into the pellet plant for conversion into high grade pellets. The project was also to serve one more important purpose. The mines in Donimalai were having certain iron-bearing rocks, namely BHQ and BHJ, which are not treated as iron ore. They are very hard in nature and contained around 40 per cent iron in them. These are dumped as wastes at the mine site. Our plan was not to lose the iron available in these materials too. We decided to set up a separate beneficiation plant for processing BHJs and BHQs, extract iron from these materials in the form of iron ore concentrate and supply it again to the pellet plant for pellet-making, so that the wastes, whether in the form of BHJ, BHQ or tailings, could be converted into saleable products. Meanwhile, NMDC's mineral-processing laboratory in Hyderabad succeeded in establishing the method of beneficiation for BHJs and BHQs by using a method of spiral gravitation so that the cost on power in extracting iron from those rocks could be minimised. With these initiatives, NMDC advanced well in the area of forward integration.

The company also took initiative in the area of backward integration. For a miner, backward integration means exploration. It means discovery of minerals in the womb of Mother Earth. In India, exploration has always been a neglected area. Reserves of iron ore in India had remained stagnant at the level of about 25 billion tonnes for a long time. About 60 years back, Australia had reserves of about half a billion tonnes of iron ore. It had grown more than 100 times over the decades only due to rampant exploration of resources. Similar things did not happen in India, mainly because the role of various agencies for carrying out exploration was not defined and as such India's resource estimation remained stagnant. The departments of mines and geology in various states were not equipped with the necessary resources to carry out exploration. We decided in NMDC that as a public sector company, we owed something to the nation. I discussed this issue with the then Cabinet Minister and wrote to the governments of mineral-rich states that NMDC would offer its free services to the state governments to carry out extensive exploration in the respective states and submit the data to the concerned authorities. Surprisingly, none of the state governments except one accepted the offer. One state government wanted to have a discussion and insisted that we should sign a legally valid undertaking that NMDC would have no claim on any of the deposits in which it would carry out exploration. It was really shocking that an honest offer from NMDC was viewed with so much suspicion. NMDC, however, mobilised its resources and started working in one state, which had accepted its offer. The initiative taken by NMDC, was however, useful in carrying out more aggressive drilling in its own mining areas. In Bailadila itself, about 600 MT of additional reserves were found by using the most sophisticated drilling machines, which the company imported. Additional drilling was also carried out in Donimalai. This mine was touted to completely exhaust its resources by 2010. However, due to the new discoveries of iron ore waste in this area, the mine continued production and is in operation till now.

As I had mentioned earlier, the main challenge for the company was to move out of its geographical area of operations and to take its output beyond 30 MT. This was difficult not only because of the problems of logistics, which restricted the production from Bailadila mines, but also because no new mines had been available to NMDC for a long time. NMDC's efforts to secure new mining leases in Odisha and Jharkhand had not succeeded. In Odisha, the company had carried out exploration work in one of the large deposits, but the mining lease was proposed to be given to a joint venture in which a majority stake was to go to a foreign company. NMDC was offered a stake of five per cent only. The company carried out exploration in another large deposit in Karnataka, but it was not given the mining lease over there. Rather, at one stage, the concerned ministry of the Central Government decided to offer the mining lease to a public sector trading company, which had a tie-up with a well-known company in the private sector, by invoking a special provision of the mining law. I will discuss it in greater length later.

In India, the iron ore mining sector is characterised by innumerable mining leases with holdings as small as two hectares and above. Some of the mines have reserves of just about 3–4 MT and the lessees do not have any experience or resources to carry out scientific mining. For such mines, as small as they are, investment on mechanisation, preservation of ore and mining on scientific basis are not possible simply because the economics of mining do not permit the same. Consequently, a large number of mines in India, especially those in the states of Odisha and Karnataka, carry out semi-mechanised mining including selective mining. The result is a rampant violation of mining laws and degradation of the environment.

When these issues were later taken up by the Honorable Supreme Court of India and it appointed committees and commissions to investigate into the matter, glaring violations of mining laws came into light and a large number of mines were directed to close their operations. While mining leases were thus distributed in random, the company which was denied leases despite its credibility was NMDC. As a public sector company, and as a large and efficient company, it should have got preference in allocation of mining leases, but it did not get it. On the contrary, in many cases, it was discriminated against. The situation possibly was better in Chhattisgarh. Barring the fact that, in 2006, the state government did deny the rightful claim of NMDC on two deposits and allotted them in favour of two private companies, the Government of Chhattisgarh consistently supported public sector companies in general and NMDC in particular. Even before we decided to set up the steel plant, a large deposit was given to NMDC on lease in a joint venture. Another deposit was also recommended for allocation as a captive mine soon after we decided to construct the plant. The state's attitude was exemplary and it gave me more and more confidence that our venturing into steelmaking was an initiative in the right direction. But then, despite the new deposits coming to NMDC, the problems on evacuation continued to exist. Within a couple of months of my joining, we signed an agreement with the railways in association with SAIL and the state government to lay a new railway line in the direction of Rawghat to establish a new route for evacuation. This initiative ultimately did not help us, because even after many years, the survey work could not be completed due to growing insurgency in the state. Nevertheless, we took actions in constructing a new mine in Bailadila and one in Karnataka, which was adjacent to our existing operations. Evacuation of material from the Karnataka mines was not a problem, but in Bailadila, the challenge was enormous. We decided that we must increase the rate of evacuation by doing whatever was possible to augment the capacity of the existing railway line. We were successful in installing a uniflow system, in persuading the railways to construct loop lines and also in providing rakes of higher capacity. All these initiatives increased the evacuation capacity by about six MT pa, which helped us evacuate additional production from the existing mines. We decided to approach the railways to lay the second railway track at least for the 150 km stretch from Bailadila to Jagdalpur. We confirmed that the entire cost of the project would be borne by NMDC. The railways agreed, but it took years before the job could be started. Doubling of the railway line between Jagdalpur and Vizag was a huge challenge and would take many years to complete. The stretch up to Jagdalpur was easier to develop and we decided to take it up first.

While dealing with the railways, my observations had been that the organisation, which was known for its efficiency, capability and intense co-ordination within the organisation had lost much of its shine over a period of time. Even 30–35 years back, it had been an example of efficiency for other organisations. In the early 1970s, when I started my career as a probationary officer in one of the major ports, the study of railway manuals was a must and these used to be referred to as the ultimate answer if there was any uncovered area in our procedural guidelines. However, when at the concluding part of my career I used to co-ordinate with the highest level of Railways, I found that the organisation had become slow and a little confused. It also lacked co-ordination, even in the highest level. When we approached the railways for doubling of the railway lines from Bailadila to Jagdalpur and agreed to bear all expenses, it took months for getting the draft MoU between the two organisations cleared, and ultimately it took more than a year for the draft to be approved by the Railway Board and that too after a lot of persuasion by the Railway Board Chairman himself. However only by coming across such experiences, we realised how important railways had been for the country's infrastructure and development of industries in the country.

While we were struggling to increase our production and find solutions to our infrastructural problems, we observed that for realising the potential of the company, it was necessary to get the existing mining leases renewed, to secure the environmental clearances for increasing production from the existing mines and to secure fresh leases. The period of lease for the Donimalai mines was due to expire. Our application for extension was under process. I decided to meet the Chief Minister and persuade him to grant extension. The Chief Minister was later subjected to a lot of criticism for almost everything he did, but I found in him an administrator who could take prompt and unbiased decisions. I have always received from him courtesy, cooperation and support. The Chief Minister took less than a minute to appreciate my request. He said that he only wanted an assurance that no iron ore extracted from the Donimalai mines would be exported out of the country, which was fine with us. I took 5 minutes time from him, prepared a handwritten letter and gave it to him. He immediately passed an order for extension of the mining lease. Later, while the papers went to the Central Government, the condition imposed by the Chief Minister regarding non-export of any material from the Donimalai mines was rejected by the Central Government, but we decided to maintain our stand because, notwithstanding the technicalities, it was our commitment to the Chief Minister not to carry out any overseas sale and, therefore, no material from the concerned mines was exported.

Meanwhile, our persuasion on increasing the environmental clearance (EC) limits for increasing the production from a number of mines in Bailadila got the approval of the concerned authority, but these were used more for availing a flexibility in our production plans than to increase production because we could not evacuate beyond a certain volume of production. Meanwhile, the recommendation of the Chhattisgarh government to grant the lease of a new deposit as the captive mine for our steel plant to a joint venture company with NMDC holding the majority stake was sent to the Central Government, but the Ministry of Mines held on to the file for months together and took no actions except for sending queries to the state government and repeating the queries few months thereafter, causing adequate frustration to our people chasing the file. A few months before the incident, I had met the Chief Minister of Jharkhand, and to my surprise had returned with a promise that one small iron ore deposit having high-quality iron ore would be given to NMDC on a 60:40 partnership between NMDC and the Jharkhand State Mineral Development Corporation. Within a few days, the required process was completed and the recommendation went to the Ministry of Mines for prior approval. This file too started gathering dust. When our officers expressed their helplessness, I met the Minister of Mines, but nothing happened. I had a distinct feeling that in some circles there was a bias against public sector companies. The situation became more agonising when we came to know of the decision of the Ministry of Mines to grant a mining lease of a large iron ore deposit in Karnataka to a public sector trading company under Section 17A(1) of MMDR (Mines and Minerals [Development and Regulation]) Act. The concerned company had no experience in mining and had a tie-up with a large private company for sharing the revenue on unequal terms. We were surprised; so was the Government of Karnataka. The latter informed the Ministry of Mines that the whole issue was under litigation and grant of mining lease at this stage to any party might amount to contempt of court. Our lawyer also wrote a similar letter to the minister stating that exercising powers under Section 17A(1) would violate the logical claims of NMDC which had carried out exploration in the concerned deposit and, therefore, it would not be an appropriate decision. It would be equally inappropriate to take further actions in a matter that was already under a judicial process. But nothing impressed the minister. I learnt that even within the ministry, senior officials objected to the decision but were overruled. The Minister of Steel wrote to the Ministry of Mines but to no avail. Ultimately, the Ministry of Mines was restrained from taking further actions under instructions from the Prime Minister's Office (PMO). This incident strengthened my feeling that somehow claims of the public sector companies and, more so, strong companies like

NMDC were getting the least priority in some quarters in Delhi. In iron ore mining, hundreds of mining leases were granted in Odisha, Karnataka and Goa in particular, but claims of NMDC were systematically ignored over the years, except for the Kumarswamy mine in Karnataka and two mines in Chhattisgarh. NMDC could not secure new mines, and when recommendations were made by the state government, as was done for two mines (one each in Jharkhand and Chhattisgarh) the response of the Central Government was neither prompt nor positive.

The way things were moving made me personally unhappy. I met the senior officers of the Ministry of Mines and was able to only extract another set of queries sent to the concerned state government. It was at this time that a parliamentary committee came to Hyderabad, and I was to depose before the committee. I submitted to the committee various initiatives in different fields that we were taking. I mentioned the initiatives being taken for acquiring overseas assets. The Chairman of the committee was visibly impressed. He commended the efforts of NMDC in securing foreign assets. I responded that I did not deserve his appreciation. The Chairman was visibly taken aback. I then explained to the Chairman that NMDC's efforts to acquire overseas assets was not an act of inspiration but that of vehement desperation. We acted in disgust, because despite our strengths, we had been denied new mining leases in our home country for years. NMDC had set examples in environmental protection, we were the role model for our work in CSR and our records in production and productivity were of global standard. We were using the best techniques required for scientific mining. We were the only mining company in India with a full set-up for research and development. We were the company which was trying to use even the lowest grade of ore to extract minerals out of it lest an ounce of natural wealth was wasted. Yet, despite all our efforts and achievements, NMDC was not considered eligible in its own country for grant of mining lease. The committee was gobsmacked. Everybody in one tone said that the committee's sympathy was fully with us. If the company needed any help from them at any stage, they would come forward to help the company. The committee said that their introduction with NMDC was an eye-opener for them. Though the committee's response gave us lots of satisfaction and encouragement, it did not get us the clearance from the Ministry of Mines.

A new minister took over the charge of Ministry of Mines after some time. I had gone to him to congratulate him for assuming the office of the minister. He was a decent man. He accepted the bouquet which I was carrying and told me to sit. I preferred to stand. I told him that I had come to him carrying the agony of the company of which I was the custodian and he was the owner. I shared my experience with the Ministry of Mines to him. I recounted to him how at every stage for months after months, the clearance for grant of mining lease to NMDC was stalled in the ministry. I said that I was moving from pillar to post as the custodian of the company to secure the mining leases which we more than deserved, and it was my owner, the Central Government, which was denying the mining leases to its own company. My emotions touched him. He ordered for a cup of tea and requested me to sit once again. He told me that, as the minister in charge, he would personally look into the matter and let me know the position. Within 15 days, I got a call from him telling me that the first file had been cleared. After another 15 days, his Secretary telephoned me to say that the second file had also been cleared. I had no words to thank him, but in the overall environment, he was more an exception than a rule.

Meanwhile, without getting much of headway in iron ore, I decided to use our potential in coal mining. I wrote a letter to the Minister of Coal suggesting that NMDC could use its resources to carry out mining operations in such coal blocks which had been discarded by Coal India. I explained that due to some sort of stagnation in production by Coal India Ltd and other companies, India was compelled to import a large quantity of coal. NMDC could contribute its resources and capabilities to participate in coal-mining in a big way, so that the domestic production could be increased and, to that extent, the import requirement could be reduced. We promised to sell coal under the same pricing formula that was being followed by Coal India Ltd. There was no answer. On the contrary, I got an indication that the coal ministry was not happy with my proposal because they had other ideas for allocating the coal mines.

> While trying to expand the activities of NMDC, I was guided by the principle that what was good for the country was good for the company. It apparently reflects a patriotic value, but it has a clear economic logic behind it.

If the country needs certain resources, it is because those resources are either scarce or the availability of it is not adequate to meet its demand. Investment in the extraction of any such resource is always preferable because of the high value associated with it. We intended to invest in coal, not only because it would be good for the nation but also because the growing demand for coal would make investment in coal mines very attractive. NMDC had unfortunately picked up some mines on commodities of lesser value, which were practically non-starters. A limestone mine had been acquired in Arki in Himachal Pradesh, which contained steel-grade limestone to cater to the requirements of the steel plants of India. However, locational disadvantages made the project unattractive as the landed cost of limestone in the steel plants located mainly in Eastern India was estimated to be much higher than the price at which limestone could be procured from other sources. We decided not to abandon the project but to slowly push it forward and to execute it at a minimum cost so that the project could get into production stage only when the price would become comparable with the supplies from other sources.

With the opening up of the global economy, steelmakers could source limestone from Oman, and the cost was lower than what could be expected for Arki limestone. The option for the steel mills was very clear, that is, to procure their raw materials from the cheapest source even if it involved import, and not from the costlier domestic sources. But then, doubt always persisted in our mind, if the material was available in the country and the extraction of the material could generate employment, income and create infrastructure, whether the determining factors should be these or the price alone.

> It is possibly necessary to weigh a decision in the context of the overall social and national interest and not the narrow commercial interest alone.

Sometime back, when the price of iron ore dipped to a level of less than $50 per tonne in the global market, the entire world expected that China would shut down most of its iron ore mines in which the cost of production itself was around $100 per tonne. In China, the grade of iron ore is low and they need to be beneficiated. But China did not close its mines, except a few, because they felt that they should not abandon their own resources for the cheaper material sourced from abroad.

> I think, this is one of the advantages of a totalitarian economy where decisions are taken on the basis of a holistic vision based on the interest of the entire economy or entire nation and are not left to be led by the force of market alone, which obviously is not always the best guide.

Almost the same situation happened in NMDC's magnesite mine in Panthal in Jammu and Kashmir. Magnesite is a very useful refractory material used largely in the process of iron- and steel-making. When the magnesite mine in Jammu and Kashmir was acquired by NMDC, there was huge enthusiasm in the area. A magnesite mine is associated with magnesite processing for production of dead-burnt magnesite, which is consumed in the production of refractory fields. Because of the fact that a process plant is required to be set up in the same location of the mine, the project is expected to require more investment and generate more employment than a stand-alone mine. However, soon after the mine was acquired, Chinese exporters of magnesite developed a free run in India and started selling magnesite at very low price, which made the Indian project totally unviable. The project remained idle causing helplessness and agony of the local people.

Fortunately, the situation started changing after a decade, and I could see in 2008–09 that the project did not look as unviable as it was in the past. But by that time, the lease expired and I was required to meet the Chief Minister to request him to extend the lease. I reached Srinagar and met the young Chief Minister of the state. His question was very unambiguous. He asked me, how the state would be benefited by extending the lease of a mine which remained undeveloped for the previous two decades. I told the Chief Minister that the situation had changed and it was possible now to start the mining activities as well as build a process plant for manufacturing dead-burnt magnesite for use in the Indian industry in particular. I added that every rupee of profit earned by NMDC from the project would be invested back to the state to develop other minerals, so that the project could have a multiplier effect on investment and employment generation in the state. It took no more than 20 seconds thereafter for the Chief Minister to take the file and sign it.

While working for NMDC, as also during my tenure in Hindustan Copper, I developed a clear conviction that contrary to the usual belief, the management of a public sector company always enjoyed more freedom, autonomy and support while taking decisions. This had not just been my experience. When I exchanged my views with many of my other colleagues, especially those who had worked in both public sector and private sector in top positions or had very close interactions with the top management of both public and private sector, they too shared the same view. I wonder whether my commitment to the Chief Minister of Karnataka that no export would be effected from the NMDC mines in Karnataka or my commitment to the Chief Minister of Jammu and Kashmir that no profit from the mines located in Jammu would be taken out of the state and the same would be reinvested in the state itself would have been possible had I worked for any private sector company in a comparable position. My experience has always been that the general perception that the public sector companies are bureaucratic, slow in decision-making and dependent on the political bosses for taking decisions is largely incorrect; rather, the opposite is true. There are exceptions, a few of which I myself was exposed to, but then, as long as there is a clear and understandable logic for taking the decision, the entire machinery will support it, unless there is any vested interest anywhere, which, at least in my experience, have been few and rare.

I cannot resist the temptation to share two of my experiences in decision-making in the public sector. One was when I was just 26 years old, working for a major port. In the port industry, the wages of workers used to be negotiated at the national level and implemented through the Ministry of Shipping and Transport. In one such pay revision exercise, three separate grades were decided for different types of operators of cargo-handling equipment. Those operating the smaller machines such as forklifts, tractors etc. were given the lowest grade, while those operating heavy cranes etc. were given the highest. I discovered in my limited wisdom that this would tantamount to keeping three sets of people for operating three different types of equipment. While that was possible in a large organisation, in a smaller organisation, it would mean that some people would always remain idle because there could be no occasion that a ship would carry or load such cargo which would require all types of equipment at a time. So I decided to limit the number of operators to the bare minimum, gave them a new nomenclature, organised training for them to learn to handle all types of equipment and gave all of them the highest grade available. It meant that the manpower came down by more than 50 per cent, utilisation was almost 100 per cent and all operators being given the highest pay scale. The wages of each worker went up by 10 per cent on an average. All these decisions, taken in the late 1970s, were not noticed by my superiors at all. When the implementation report was sent, the ministry immediately reacted and asked for the reasons for giving the highest scale to all the operators by putting all of them under a non-standard job title. My boss called me and showed me the query from the ministry. I explained. He said that whatever I told him should be written and clarified to the ministry. Within a few days, we discovered a circular from the ministry advising all ports to follow the same system adopted by the port I was working in. It gave me the confidence that if one works on logic and without any hidden agenda, there was no reason to be victimised. Of course, another story that I recall gave an impression which was a little different. Having just joined a large public sector company as a director of HR, I noticed that in a promotion exercises, the senior-most man who was also known for his capabilities was not recommended for promotion. I was told that the promotion rules were very specific. If a person did not get at least 'very good' ratings consecutively for the previous three years, he/she would not be considered for promotion, notwithstanding his/her other qualities. The concerned man was rated 'excellent' for two consecutive years. But on the third year, he was rated 'good' by his reviewing officer, though he was rated 'excellent' by his reporting officer and therefore, he was left out. He was to wait for another three years to be eligible for promotion provided that on all these three years his performance appraisal rated him as 'very good' or 'excellent'. I was surprised. I found that it was totally irrational. I explained the position in a note and proposed to CMD that the person concerned, if otherwise suitable, should be promoted. I also said that very soon we must change the rule to avoid such situations in future. The Chairman agreed with me. The gentleman was promoted. I almost forgot the matter thereafter. One day, an officer of my department came to me and told me that my confirmation was stopped by the ministry on the ground that there was a complaint against me on violation of rules and giving undue favour to a person in granting him unlawful promotion. I was really amused. I was then told that the file was moving from one senior officer's table to another and there were many exchanges of notes. I decided not to do anything and simply ignore it. After a delay of about six months, my confirmation papers came and I came to know that at last good sense had prevailed at a very senior level and the file was closed. It again made me believe that there could be some people somewhere who may create problems, but ultimately nothing can happen if the decisions are taken in the best interest of the organisation and the people. It is not, however, always true that all initiatives ultimately lead to success or even that the most honest initiatives are not faced with resistance.

While in NMDC, I was concerned about the fate of another large mining company called Kudremukh Iron Ore Company Ltd (KIOCL). At one time, it was one of the largest mining companies of India. It was involved in the mining of iron ore in the western hills of Karnataka, beneficiating and pelletising it and exporting it to Iran. In 2005, the entire operation of the company was put to halt on grounds of environmental protection. The mines of KIOCL are situated in the Western Ghats, which are known for their biodiversity and sensitive ecology. The issue relating to the propriety of carrying out mining operations in Western Ghats was taken up by the Supreme Court, which had directed KIOCL to wind up its mining operations. From 2005, all the operations of KIOCL, known to be a vibrant organisation, came to a halt and the company survived only by operating its pellet plant by purchasing iron ore from other sources. Despite being fully convinced that no human activity should be allowed to destroy or adversely impact the flora and fauna of an extremely sensitive environment, I felt always hurt about the plight of KIOCL. In any country, and more so in a country like India which needs more economic activities to take care of its growing population and its need for wealth generation at a consistent rate, it is necessary that its mineral resources are extracted and utilised for the benefit of the country. Closure of mines in a company like KIOCL, which not only had been fundamentally a strong mining company but also had excellent contributions in the field of environmental protection, was a blow to the Indian mining industry. I had the opportunity to visit one of the very few underground iron ore mines, namely the Luossavaara-Kiirunavaara Aktiebolag's (LKAB) Kiruna mines in Sweden. Kiruna is a government-owned mine, which is the largest underground iron ore mine in the world. It operates at various levels beneath the surface and employs about 1,800 people including 400 in the mine. For any iron ore miner, a visit to Kiruna is almost like a pilgrimage. Having returned from Kiruna, we decided to put up a proposal to the Government of Karnataka to work for converting KIOCL into an underground mine. We felt that a joint petition from all concerned proposing that the mining operations could be carried out without impacting the surface, and thereby protecting the surface cover and the ecology, could be an option which the Supreme Court could consider. It was also felt that change of management of the company could give the right message to the Supreme Court. NMDC was already known for its work on protection of environment. While reopening the Panna diamond mine, the Supreme Court had made an exception to allow the mining operations to be carried out by NMDC inside a reserved forest for the first time because it had faith that NMDC would do its best to ensure that mining operations would not have a conflict with environmental protection. We, therefore, decided to acquire the majority stake of KIOCL and approach the court with the support of the state and Central Government to allow the opening of KIOCL mines with the condition that mining operations would be restricted to underground mining only so that the surface remained unaffected. To me, it sounded like a good idea. The Karnataka government was enthusiastic, as was the Central Government. However, there were reservations from the top management of KIOCL. They failed to understand how the change in management would make the Supreme Court change its mind. The very fact that the Supreme Court not only was to be convinced that underground mining was better option but also needed to be assured that a tested management would implement the scheme could not convince KIOCL management. They got support from the status quoists in the government and our dream to work for reviving KIOCL remained unrealised.

In the meantime, the Government of India came out with a proposal to thoroughly modify the MMDR Act to the extent that they wanted practically a new Act to replace the old Act. Many provisions of the proposed Act appeared to be unrealistic to us. The approach also appeared to be short-sighted. The mining assets in India were clearly opened up for acquisition by overseas giants who could participate in auctions and had every possibility of winning the bids, because the exchange rates of most of the international currencies were much higher than the purchase power parity of the concerned currency vis-à-vis Indian rupee. There was no reservation for the public sector. The government wanted to have a one-time value extraction through the auction of mines instead of allowing the mines to be developed by the public sector companies, especially a strong company like NMDC, which could give the government an opportunity to have perpetual value extraction from the mines allotted to such companies.

Contribution by the mining companies of an amount equivalent to the royalty paid to the district development fund was an impractical idea. When the mines have high margins, the money diverted to the district development fund is so high that it is impossible for the concerned body to utilise the money. Then again, when the mineral price is low, in addition to paying royalty, which is quite high, payment of a similar amount to the district development fund makes the mining operations unviable. An Act is a long-term arrangement, and when such provisions are incorporated in an Act, the industry suffers as the situation undergoes changes. I took a very strong view against a number of provisions of this Act. Fortunately, the Mines Minister himself put his foot down and the Act was not allowed to be passed. Later, when the new government came in, an ordinance and later a bill were passed modifying the existing Act with a lot of sensibility and pragmatism.

When I look back in retrospect, I feel that despite innumerable obstacles and hurdles, our struggle to expand our mining business in NMDC, to increase our resource base, to diversify into different minerals and to organise backward and forward integration, all produced quantifiable results. Often, we used to feel that we were not getting success despite our dedicated efforts, but in the end, when we took stock of the final outcome of the efforts we all had put in, we felt that we had no reason to feel frustrated. NMDC's production capabilities and resource base increased substantially during these years. Apart from increasing the capacities of the existing mines in Bailadila, we were able to almost complete the construction of a new mechanised mine to add to the production from Bailadila sector. Two new mines came into our fold in Bailadila sector itself though under a joint venture arrangement with majority stake being held by NMDC. Construction of a newly acquired mine in Karnataka in the same area where our Donimalai mine was situated also commenced to make the capacity of Donimalai sector twofold. The stage was set to take the company at a production level of 50 MT of iron ore pa despite taking into consideration the problems of logistics, which were handled with patience and consistency. Even though the Railways had only a single line connection between Bailadila and Vizag, the capacity of the existing infrastructure was augmented by about 40 per cent, mainly by introducing a uniflow system, which provided for the facility for the rakes to move forward after loading without requiring them to be drawn backwards and parked in the railways yards. The Railways provided for loop lines almost in each halt in order to ensure uninterrupted movement of the rakes. They introduced new signal system and new rakes for carrying about 10 per cent higher load. All these helped, though occasional bandhs and disruptions continued to restrict the use of full capacity by the railways. We continued explorations in our own mines both in Bailadila and Donimalai sectors and discovered additional resources of more than 600 MT. We got the Central Government approval not only for the captive mine for our steel plant in Chhattisgarh but also for an additional mine in Jharkhand again under a joint venture structure. We were successful in persuading the Railways to start work for doubling the track in Bailadila–Jagdalpur sector, which covered about one-third of the total distance from Bailadila to Vizag. Once this construction is over, this will facilitate not only additional evacuation from Bailadila but also transportation of iron ore from our captive deposit in Bailadila to the Steel Plant. The beneficiation plants and the pellet plants that we decided to set up opened up new dimensions on preservation of natural resources by converting wastes into marketable products.

NMDC was no longer a passive player in the process of decision-making and policy formulation at the national level. The company had not only been successful in preventing wrongful allocation of a mine by the highest level in the concerned ministry, it actively participated in deliberations on the new mining bill. Its opinions were not only aired but also heard. NMDC was no longer a sleeping tiger. It came to know of its strength, it succeeded in establishing its rights, it was recognised as an important player in Indian mining sector and it became known globally. NMDC started getting offers from Australia, Brazil, USA, Canada and Russia for acquisition of mining assets in those countries. Some of the offers came from very important quarters. NMDC was guided by two factors only: national interests and its own business interests. It took over a mining company in Australia and became the first public sector company in India to have a successful overseas acquisition. It discarded many offers after extremely judicious scrutiny of the concerned assets. In retrospect, it is evident that those were the right decisions. In NMDC, we created a culture of objectivity, transparency and speed. NMDC became an aggressive company, successful in pursuing its cause at every level. It helped us materially, but may be more importantly, it improved the morale in the company, the degree of professionalism and confidence and dignity, which the people of NMDC came to be known for. NMDC no longer remained the sleeping tiger. It became a force to be reckoned with.

6

MANAGING PEOPLE

One of the fundamental strengths of NMDC had been its motivated workforce and constructive unions. The officers were hardworking and the overall work atmosphere was positive, supportive and inspiring. The employees were content and modest. These were sufficient for the company to survive and grow in its own stream. This is exactly what had been happening for many years in NMDC. It had been a stable company. The company knew that it could construct modern mines and run them efficiently. It had mines with iron ore of one of the world's best qualities, which had ready markets in India and abroad. When I joined NMDC in 2007 and introduced some new visions into the future of the company, things no longer remained as they had been for many years in the past. For some, it was an opportunity; for others, it was a threat. As the doors and windows of the company started opening up and winds flew in from outside, not only smooth breeze but also dusty storms found their way into the company. A section of employees felt that these changes were unnecessary and to some extent unwelcome. Some, however, felt that this was what the company should have welcomed even earlier. Resource management acquired a new dimension as new challenges started coming up. Higher profits and higher investments by the company inspired the workers to demand higher remunerations in line with whatever was prevailing in other companies of the country. They no longer remained happy with whatever they were being paid. The unions became more active, and with this, inter- and intra-union tensions became visible. Officers aspired for higher growth and became more concerned about their rights and more aware about the opportunities available to them outside the company.

So far as I am concerned, I was not surprised with these developments. These were inevitable consequences of growth and diversification. My effort was to channelise these sentiments into constructive participation of employees in the drive of the company to grow. NMDC already did have a structure of federation of unions, which represented all the unions jointly in most of the participative forums and for collective bargaining. There were a number of active unions in the company. Almost all central trade unions had representatives. There were unions affiliated to AITUC, INTUC and BMS. However, they were together in the federation sharing different portfolios. If the President was from one union, the General Secretary was from another and the Vice-president was from yet another union. The federation became more active with the changes happening around. Senior trade union leaders of national level were the top office bearers of the federation. We knew that they had always provided constructive leadership to the federation, but now they had to handle different levels of aspiration of the workers they represented. Their task was difficult. We decided to be completely transparent with them, to explain to them all our plans and initiatives and to take their views on everything we were planning to do. We knew that transparency had no substitute and honesty was irreplaceable. The mature leadership of the federation understood us very well. They understood the sentiments of the workers fully and they provided the right leadership at the right time.

Many a times, they did not hesitate to rebuke the workers for their excessive demands. They were never hesitant in arguing for the right cause and they were not shy of loudly supporting the management whenever it was necessary. Such a constructive body could be an asset for any company, and many a times I feel that much of our initiatives in taking NMDC out of its own cell would not have been possible but for the support we received from the federation. Let me make a special mention in this connection of the President of the union, who had for many years been a leftist parliamentarian known for his long struggle for the cause of the workers. He was widely respected, but he never encouraged any demand of the workers coming out of greed and not based on genuine logic and reasoning. He never hesitated in criticising the management bitterly in case it was found at fault, but never showed any disrespect to anyone at a personal level. He was accepted as a leader by the leaders of all the constituents of federation and, therefore, though there had been some disunity amongst the union at a local level or tensions within the unions at some level, there was no disharmony amongst the union leadership in the federation. After almost four-and-a-half years of my joining NMDC, when I retired from the company, to my surprise, this veteran Member of Parliament had organised a farewell for me on behalf of the federation. I was very surprised because he was never known for having shown such gestures in his life. In fact, he admitted this fact at the beginning of its emotional speech, which he had delivered during the farewell. He said that this was the first time that he had organised a farewell for a person who was the head of the management, but he was doing so because the person being given farewell to was just different. In my drive to make NMDC a great company, I always got support and encouragement from him, but I never expected such kind words.

In a large company like NMDC, especially in the phase of growth and diversification, active support of the unions was very necessary. In Hindustan Copper, I had experienced how union support could be an essential prerequisite for the revival of a sick company. In NMDC, a culture of co-operation between the union and the management was already existing in the company, but that was never tested in the context of rapid diversification programmes initiated by the management. In fact, resistance to change is a common phenomenon in management. It is also reflected in any union–management relationship. Not only the workers and their unions but also most of the executives feel insecure when a company ventures into new terrains, in the field of business as also in its social relationship. While I was preparing to leverage the fundamental strengths of the company for taking it forward to a higher plane of growth, I was aware of the need to handle this aspect. From my experience in other organisations, I was confident that what was necessary was to communicate, as widely and as intensively as possible. Most of the resistance to change occurs from distrust, which again is the outcome of communication gap. I decided that we must communicate not only to the unions but also to the workers directly as well as to the executives and share with them everything that we were planning to do, including the problems that we were facing and apprehensions that we were carrying. We were interested not only to communicate but also to get feedbacks, many of which proved to be very useful in the future. While communicating with employees, we spoke not only about the company and its plans but also about other issues including the macro-issues such as Indian and global economy, growth prospects, market, and technology etc. We also encouraged the employees to understand and participate in every area of the functioning of the company, keeping their heads high and acting without fear. When we decided to set up the integrated steel plant, we got a very honest response from our employees at all levels. Initially, many of our employees were apprehensive. Many officers, who had spent a large part of their working lives in the mines, felt that with the coming up of a steel plant, the company would be converted into a steel-maker from a miner and that their importance/value in the company would recede. Some felt that, with new people coming to manage the steel plant, their opportunity to grow to the higher positions would be adversely affected. Workers felt that those in the mining units would be numerically overtaken by those in the steel plant and, therefore, their say and positions in the unions would go down. There were proposals that the steel plant should be developed as a subsidiary and not as a part of the company. All these issues were deliberated upon at length and our decision, which had been agreed to by all concerned, was that the steel plant should be an integral part of the company. The employees of the company, except those who were specialists in mining operations or exploration or mine development, got immediate opportunity to work in the steel plant and avail of better growth prospects. Those associated with mining, due to their own specialisation, did not have any conflict with those specialised in steel making. More positions were available at the top. More employment opportunities were created, which were found attractive to the employees, especially those from the neighbouring area, for their children to avail. Above all, the employees were convinced that what was happening was good for them and for the stability and growth of the company they belonged to. Unions supported the idea from the conviction that it would bring prosperity not only to the company but also to the community, which their top leaders, being leading politicians, were politically and morally committed to. When we decided to expand our activities on CSR, the unions and the workmen came forward to actively participate in many such initiatives, which I would be describing in later part of this book.

As I have observed, in all the organisations I have worked in, the human mind is essentially creative. However, human beings are excellent in suppressing their creativity or using it selectively. The job of the management is to unleash the creative potential of the employees by creating an appropriate structure and environment within the organisation. In a large organisation, employees are managed collectively, but the human mind is too delicate to be treated as a part of a group, especially if the group is large. Unions may settle the economic grievances of the employees but, at the core, employees need personal recognition. It is difficult in a large organisation to create a structure in which people's own ambitions, own aspirations and own sentiments will be taken care of and they will be directly and personally encouraged to come out with creative ideas, which will not only be the source of motivation for the person concerned but also be a source of value for the company. I have seen that in small organisations, where there is direct relationship between the owner and his/her employees, even though the service conditions are much worse, the degree of motivation is much higher as compared with what we observe in a large organisation where not the owner but the structure decides the relationship of an employee with the organisation. Having gone into this issue deeply, I have come to the conclusion that in a large organisation, it is not really possible to completely do away with the structure. However, large organisations can be broken down into teams where the head of the team will do the same job as that of the owner of a small organisation if he/she is empowered to do that. A survey carried out sometime back indicated that in a large organisation, one of the most important factors which motivated the workers was respect and love for and attachment with the immediate superior, who treated them not as faceless entities but as living individuals, each distinctly different from the others. I tried to follow this principle in NMDC.

For unleashing the creative potential of employees, quality circles play a very important role. With more and more contract workers being deployed in manufacturing work nowadays, the concept of quality circle has lost much of its relevance. However, in a company like NMDC, where not even a single contract labour is engaged in operations, quality circle continues to play an important role. In fact, NMDC had quality circles since long, and quite a number of them not only existed but also played a very useful role in extracting the best talent from their members. As is well-known, quality circles are small groups of employees which are voluntarily formed to jointly discuss, analyse and assess the existing systems, procedures, technologies and processes in order to find ways and means to bring improvement upon the same. I was very proud to see the quality circles operating in NMDC with quite a bit of effectiveness. I decided to encourage the movement to gather strength and to not only become more relevant but also be recognised to be so within the organisation. Some of the quality circles were invited to make presentations before the top management. I was personally present during the presentations and was thrilled to see the analytical approach exhibited by the members, identification of problems and assessment of the choices available to the team to solve the problems or to come out with altogether new ideas, which could add quantifiable value to the company. Our teams were encouraged to make presentations at the national level and also at the international level. Some of the teams were sent to Japan and earned laurels not only for themselves but also for the company.

Due to intense competition, every company, whichever sector it is in, strives to cut down cost, improve quality and reduce the delivery time. This require every process to be reviewed, every activity to be assessed to determine its relevance and every system to be questioned to assess the degree of redundancy associated with it. Whenever these initiatives are taken at the top, most of the times, the situation at the ground level are not taken into consideration. Micro-aspects are ignored and decisions are taken at the macro-level, which may not offer the ideal solution to the problem. Quality circles provide an opportunity to an organisation to understand every relevant aspect at the micro-level and allow the talents at ground level to come out with solutions. Quality circles are great motivators for its members.

The members not only derive inspiration from the success they achieve, they also come out of the drudgery of routine work and get an opportunity to exhibit their intellectual skill. I found that quality circles gave back to the organisation much more than the support they get from the same.

Another problem which occurs in an organisation is the lack of day-to-day evaluation of the performance of the employees, especially the executives. The system of annual performance appraisal is more a formality than an actual assessment of performance. It is quite unacceptable that an employee's performance should be assessed once in a year and that the superior will spend not more than a few minutes to fill up the forms designed for the purpose. First, it is not justice done to the employee concerned and second, whenever assessment of performance is done in such a manner, the purpose of assessment is defeated to a large extent. No bell curve system or no system of open assessment can improve the situation. In fact, the employees are not even sure that they are even watched. They sometimes doubt whether their existence is even noticed by the very superiors who they are supposed to be reporting to. As the organisations grow larger, the alienation of an employee from the company grows faster. I decided that every executive, wherever he/she might be working, must maintain two sets of diaries provided to them by the company. Every day, at the end of the working hours, every officer was required to briefly write down in just 8–10 sentences, what he/she had planned to do during the day, what did he/she do, what were the obstacles, if any, and how did he/she overcome them. At the end of every week, the diary would go to his/her superior and during the next week he/she would use the second diary. During the course of the next week, he/she would get the first diary back with the comments of his/her superior. The comments would include guidance, appreciation and suggestions for improvement. Immediately after the system was introduced, there was huge resistance from every level. Most of the executives discovered that they had nothing much to write. The superiors were embarrassed, because they always demanded more people, but now they found that they did not have enough work to keep their people engaged. I decided to be firm in getting the system introduced. I directed that nobody would be promoted without the diaries being presented to the selection committee and without the selection committee being satisfied with the work performed by the concerned officers during the previous period. There was no option for the people but to start writing the diary and soon we discovered how easy it was to understand the role played by each executive in the organisation. The senior officers had also to be alert in writing the annual performance appraisal reports because the top management at random used to check the annual performance appraisal report and its compatibility with the diaries maintained by the officers concerned. It was a system, which I found very difficult to implement, but I was determined and ultimately it became a part of day-to-day work for the executives, most of whom, though reluctantly, finally accepted it.

Another system which I decided to introduce was to dispense with the usual system of personal interviews conducted by the departmental promotion committees for considering promotions of an executive to the immediately above level. Many a times such interviews become mere formalities and meaningless wastage of time. The officers who appear in the interviews also take it casually knowing that the prospect of their promotion does not depend much on their performance in the interviews. I directed that each officer appearing in a promotion interview would have to make a brief presentation on any relevant topic of his/her choice before the interview board. The presentation would cover some part of the assignment the officer concerned is handling. The officer will explain how much he/she could understand his/her role in carrying out the assignment, how well in his/her own assessment could he/she handle the assignment, what were the problems, if any, and how did he/she tackle them. The officer was free to describe his/her work in whichever way he/she felt right, and the interview board's role was limited to asking for clarifications, if any, on the presentation made by the officer concerned. This introduced a high degree of objectivity in the system. At the end of the interview, the interviewee himself could understand how good or bad his/her performance had been. The members of the interview boards were advised to assess an executive on the basis of how he/she handled his/her assignments and not on the basis of the quality of presentation material and skill to articulate.

These were small steps, but all these were aimed to bring a work orientation in the whole organisation. The performance of the workers was already assessed through a well-devised incentive system; there was nothing to directly assess the performance of the officers on a day-to-day basis. The small and effective systems introduced in the company brought about a sea change in the culture. At least it was a beginning in that direction.

Human resource management is always an extremely delicate area. It cannot be handled only through systems and procedures, and rules and regulations. Employees' behaviour, attitude and approach to work, response to situations and acceptance of circumstances are all deeply embedded in the psychological and sociopolitical context of his/her personality and upbringing. Having been in the field of human resource management for many years, I have learnt to respect certain inner sentiments of people in our country. I followed these lessons everywhere including NMDC. The first of these had been that a person was highly protective and possibly quite possessive about his family, his spouse and children. Even the poorest man in our country is the master in his own family, and he does not like to allow anybody to interfere in that. He does not like that anyone from his workplace, be they his immediate superior or a topmost executive, should try to guide and educate him on how to manage his family. An average person in India enjoys the way he is treated as a king in his family, the way he is served food and the way his words are treated to be the ultimate in his family. Maybe Congress would not have been routed from power in 1977, had it not tried to implement the family planning programme through force. This family sentiment possibly also explains why most of the industrial unrests in Indian organisations start from the canteen where, in general, the quality of food served is better than what is eaten by an average Indian in his/her own home, but where there is no personal care with which a person is given his/her plate to eat from. In a company, therefore, care should always be taken to respect this sentiment of the working people and not to take any action which may amount to intrusion into the privacy of his/her family.

Second, for an average Indian person, religion is an integral part of the culture he/she grows up in. One may not look like a religious person from his/her apparent behaviour, but when the question of religious commitment comes, such as attending some religious ceremony or even a social ceremony like marriage, the person may be very determined to attend such ceremonies and will feel hurt if he/she is denied an opportunity to do so. In our country, religious ceremonies are intermingled with social obligations and family commitments. All taken together, it becomes unthinkable for a person to remain absent from such ceremonies, and as an employer one should always avoid creating such situations.

Third, an employee always expects to be known and recognised as an individual who has a name and a separate existence. He/she doesn't like to be treated as a number or as a part of a group. He/she feels insulted, though secretly, if his/her holistic personality is not recognised and he/she is treated as a working hand. It may be that our traditional system of calling workers as skilled hands, unskilled hands or leading hands was and is counter-productive. Once the management recognises this aspect and treats an employee as a complete person with all his/her emotions, sentiments, aspirations and intellectual abilities and does not identify him with his/her hands and skills alone, the best work culture is created.

These lessons and their right implementation can make a big difference in an organisation, and NMDC was no exception.

Years have passed by since I left NMDC. I had inherited a strong company from my predecessors. The company's only weakness had been that it did not know how strong it was. It did not know that it had a much larger role to play at national and international levels. It did not know that it had the potential to do so. The people of the company were content that what they were doing was enough, what they were performing was adequate and what they were achieving was the best. Despite its strength, the company was not moving forward with the advancement of the world. In the process, the company was losing its relative strength, which most of the employees of the company were not even aware of. The world market of iron ore was going to witness a boom, an intense competition for grabbing iron ore resources was going to start within the country and globally, value addition became the need of the day and social resistance to the commercial enterprises was becoming imminent. The state was succumbing to the pressure from different interested circles and started taking the path of least resistance, which was to make the road to progress thornier. In NMDC, the people had the full potential to handle these challenges and to take the company forward through determined efforts. Man-management came easily to me, and I used it fully to inspire the employees at every level, to communicate with them, to watch them, to assess them and, finally, to empower them to take the company forward. After due deliberations at every level, when we decided to set up the integrated steel plant, 120 km away from our mines at Bailadila in Chhattisgarh, on the day of laying the foundation stone, the employees from the mines came to witness the ceremony in hundreds and thousands to share the joy and pride of being a part of the new project, which was a great leap forward in the life of the company.

7

COMMUNITY WORK

Many years have passed since Pandit Jawaharlal Nehru described the public sector organisations in India as the temples of modern India. Over a period of time, the concept of public sector has undergone erosion in value. Some of the large companies created in the 1950s and the 1960s, soon after the country got independence and adopted the principle of planned economy, have ended up as sick entities and become burdens on the national exchequer. Investment in the public sector was made with the hope that it would generate profit, which would in turn provide resources to the nation for further investments for nation-building. Public sector companies were to act as growth centres creating employment and wealth to bring in development and reduce regional disparities. There were some successes in achieving these objectives, but these were not up to the level that the nation had expected from the public sector.

To me, working for the public sector always meant working for the nation. I felt that it was an opportunity to serve the nation, the society and the community in the best possible way. I always felt myself self-motivated when I worked for public sector organisations. In my day-to-day work, I always felt that the benefit of what I was doing was directly going to the society at large. I never felt that there was any conflict between the objectives of profit-making and the goal of working for the society. A public sector company is a commercial entity, which must find its relevance by carrying out meaningful business and making profit through optimisation of cost, improvement of quality and expansion of operations. There is no scope of doing any compromise in these areas. However, the profits of the public sector do not go to an individual; it goes to the state in the form of taxes and dividends, it comes back to the organisation in the form of reserves and surpluses and it is channelised to the society in the form of work for the community in order to provide better quality of life to the people. A public sector company which does not make profit, loses its significance as a commercial organisation, but a public sector company which does not work for the society and the community, loses its moral value because it owes its existence to every citizen of the nation which has created it.

NMDC was set up in 1958, and its first set of mines were developed in the remote forest areas in Bastar in the state of Madhya Pradesh (then undivided). The area was mostly inhabited by tribal people who for centuries used to live in the region and found the means of livelihood from the forests, which they protected as their wealth. For centuries, they maintained their separate identity, separate way of life and separate culture from the rest of the country. When, after independence, the nation strove to move forward by developing new industries, new mines and new markets, there was intrusion into their domain for the first time after many centuries. The hills of Bailadila, which conserved high-grade hematite iron ore for millions of years, suddenly witnessed huge activities hitherto unknown to the people of this area. People flowed in from different parts of the country, some from other parts of the world, to work together to build a set of modern mines for extraction of valuable iron ore resources from the womb of mother earth. In the initial years, the reaction of tribal people had been a mixed one. The activities created opportunities for employment, but it also created some social unrest. However, there was no firm resistance and there was no serious problem for NMDC to complete its construction work. When the mines started operations, a large number of families from outside came to live in the area. New townships came up with schools, hospitals, parks, entertainment centres and other amenities which people living in a modern society expect to have. The gap between the local people and those in NMDC suddenly became wider and more visible. NMDC did not ignore this reality and came to support the community to the extent possible by providing training and employment, constructing roads and bridges, building schools and community centres, and providing facilities for education and health care in and around the townships of NMDC. People recognised these initiatives and accepted them with gratitude. However, as time progressed, some of these work lost initial impact.

At the time of my joining NMDC in 2007, the entire area was politically sensitive. Communist extremists or Naxalites, as they are generally called, developed their strongholds in that area. Politically and socially, the extremists created a strong base in Bastar. They took almost total control in most of the villages. They were fully armed, but unlike many other militant outfits, they used arms very selectively, mainly against those opposing them and the armed security force. There was a general sense of fear all around. In 2004, the Naxalite movement gathered further strength after the merger of two important Maoist groups, namely the People's War Group and the Maoist Communist Centre (MCC). The two groups jointly formed a new entity, the Communist Party of India (Maoist), and built their base in various states including Chhattisgarh, Bihar, Jharkhand, Maharashtra, Odisha, Andhra Pradesh and West Bengal. In Chhattisgarh, the state government promoted a counter-insurgency group called the Salwa Judum, which constituted those in Naxalite-controlled villages who opposed Naxalism and were prepared to confront them. They were accommodated in more protected areas, were armed and were engaged to fight with the Naxalites. The creation of Salwa Judum generated more tensions in the area. There were killings and counter-killings. Many lives were lost and ordinary people suffered the most. One thing was clearly understood. Naxalite movement or the movement initiated by the CPI (Maoists) was clearly a political movement. The insurgents wanted to capture power by using arms and fighting against the state power. There were systematic attacks against police and paramilitary personnel. Salwa Judum suffered the most. The entire area became tense. The villages became inaccessible. Violence became common and logic was a casualty. In such situations, organised opposition and resistance to any new initiative becomes very common. It becomes difficult to carry out organised economic activities in the area, because the organisations can neither confront nor co-operate with the armed insurgents. Some of the companies in the private sector, which obtained some mineral concessions in the area, faced steep resistance and were forced to refrain from carrying out any meaningful activity. NMDC, which already existed in the area for the previous 50 years, could continue, but was not free from fear. In 2006, armed insurgents entered the mining area and caused damage to the conveyor system which resulted in stoppage of mining operations for a few days. The security arrangements in the mines of NMDC were very elaborate. The mines were not only completely fenced, there were watchtowers in between, manned by heavily armed personnel carrying not only the most sophisticated weapons but also night vision binoculars and all sorts of protective gear. There were mobile groups as well, which were responsible for checking every possible entry point all through the day and night. Even then, the entry of armed insurgents did take place and the company suffered losses. Despite all the community work which the company had been carrying out for many years, the armed people had no problem in crossing the nearby villages and attacking NMDC. The situation in the region was difficult, and the only consolation was that NMDC was the least affected out of all those who intended to carry out mining operations in the region. The people in the region were poor and many of them started believing that companies like NMDC no longer remained useful for them. Despite NMDC being there, there was not much of relief for them. They continued to suffer from hunger, ignorance and diseases. There was some sympathy for the Naxalites who after all were supposed to be fighting for them and were their own men.

When I joined NMDC, I found that the biggest challenge for the company, even before we could talk about expansion and growth, was to ensure stability, to provide confidence to our colleagues in the mines that their lives were safe, to ensure that we could continue our operations uninterruptedly and to do that not only by providing more security but by working for the community, interacting with the people, understanding their needs and working for improving the quality of their lives. I felt that if we go and work for the community with the purpose of buying security for our people, it would be of no use. We needed to be genuinely concerned, to help the people and to empower them through our support because, and only because, we owed it to them.

I remember having made a public speech clarifying our role in the area. I said,

> "Here in the area where we are working there are people who use arms for pursuing an ideology that they believe in. We do not have arms. We may not subscribe to their ideology, but beyond arms and ideology, there is a huge space available

for us to work in. This relates to health care, this relates to education and this relates to empowerment through generation of employment and creating opportunities for earning. I am sure that there is no scope of conflict in these areas."

This defined our approach to a large extent. While taking stock of the situation, I observed that the community work through CSR had to bring direct benefit to as many people as possible. We decided to first identify the villages which surrounded our mining establishment in Bailadila. There were 58 such villages. We decided to intensify our work in these villages. Our active participation in the developmental work in these villages touched almost every part of the lives of the people living there. We commissioned mobile medical vans, fully equipped with essential medical equipment and medicines. Each van was manned by a doctor and paramedical staff. The vans would go to each village on the appointed day and time and provide free medical treatment to the villagers. The doctors would keep full medical records of the patients so that they could have a holistic picture about the condition of each patient and his/her past history. Arrangements were made to pick up the critical patients along with their escorts, take them to the company's super-speciality hospital in Bailadila and provide free treatment to the patient and food and accommodation to the escorts. This system already existed, but was further streamlined with a lot of empathy and planning. Some patients could not reach us or inform us in case of serious sickness because there was no accessibility in some of the villages. We provided mobile phones to the village chiefs so that they could inform the hospital immediately and shifting of the patient to the hospital could be arranged. Our work in the villages included education to every child, small irrigation and training on farming and preventive health care, which interestingly included distribution of mosquito curtains to prevent malaria and sinking of tube wells to prevent waterborne diseases. We wanted to build around the mines a ring of people who will have a stake with us. They would be supportive and sympathetic to us and would not provide moral support to anybody who could harm us in any way. We had regular interactions with the villagers and in some such discussions I myself had participated. The villagers had many grievances and their expectations were high. We understood their sentiments and dealt with their problems with a lot of patience. Often the villagers came and met us along with their local leaders. Some of them were known sympathisers of Naxalites. We used every opportunity to convey to everybody our intentions to reach out to people and our commitment to do developmental work in those villages. These communications turned out to be very useful and the response of the people of nearby villages was not only positive, it was clearly visible to be so.

Internally, however, I had a dilemma. We were surely working for the community. Our work was certainly helping the people, but it was not free from our vested interests. We wanted to help the community because in our heart of hearts we knew that this is what we were required to do: to provide for the safety and security of our mines and the people working in those mines and their family members. What we did was the implementation of a strategic decision and not so much genuine community service, which we were otherwise committed to do. Further, for carrying out fieldwork in these villages, we engaged two non-governmental organisations (NGOs), one of which was run exclusively by women. This was necessary because the work carried out in the villages was so intensive that a full-time presence of people to organise the work and to monitor it keeping in view the needs of the people was absolutely necessary. This again was not to my satisfaction. I always felt that CSR work was too important to be offloaded to any other agency. An organisation and its people are integrated with the community through CSR work. If the job is done through an intermediary, whatever skill they may have, the moral benefit of CSR work does not fully come to the organisation and its employees. When we planned to organise CSR work beyond these 58 villages, we did it on our own, by our own employees who voluntarily participated in CSR work.

As indicated earlier, my firm belief had been that the benefit of CSR work should go directly to the people and as many people as possible should get the benefit. Many politicians approached us for CSR work. Some proposed the construction of a bridge, some suggested the widening of a road and some felt that we should spend CSR money on beautification. I was not impressed. Having observed the situation in Bastar, my feeling, had been that ultimately we will have to create employment for improving the quality of life of people. Employment would come to a large extent from our steel plant, but agriculture and transportation etc. were also to be developed to create opportunities for people to earn. Equally important was education, which would make the young people suitable for jobs and capable of earning their livelihood.

> Education has always been the strongest source of empowerment. I felt that a closed society, which remained alienated from the outside world for centuries, should be opened up through education.

We decided to introduce a scholarship scheme to encourage students to pursue their studies and prevent its discontinuation. In that region, due to poverty, many students drop out of school before getting into higher classes even at the secondary level. Many of them are compelled to do so due to family pressure. We decided to grant scholarships to as many as 11,000 boys and girls in 330 schools covering Bastar division as a whole. Scholarships were given to everybody who secured at least 45 per cent marks in their school examinations and if their score fell short of the minimum requirement, they were encouraged to secure 45 per cent in the next test and be eligible for scholarship retrospectively. We opened post office savings bank accounts for each student and taught them to handle it. We had appointed a group of volunteers comprising male and female employees of NMDC, mostly belonging to the same area, who would visit the schools at regular intervals. The scholarship scheme provided them access to the schools. They went not only to the schools but also to the houses of students who were not attending classes. For the first time, after many years, the remotest villages and the schools situated in those villages received outsiders with open arms. The warmth with which they were received was unbelievable. Some of the volunteers could not check their emotions when they described their experience to me. When they got access into schools, they got an opportunity to assess the needs of the schools, many of which had very poor infrastructure. They provided books to the students and furniture to the classrooms. Apart from textbooks, we decided to distribute books on popular science and general knowledge so that the students could get a glimpse of what was happening around them in the world. We organised interschool competitions on sports and cultural activities. We also organised excursions of students and teachers to other places including Hyderabad. On Republic Day, in NMDC's headquarters at Hyderabad, a cultural performance by the students of different schools from Bastar became a regular feature. They stayed in Hyderabad for a few days being guided by our volunteers and interacted with the local people freely. We wanted the doors and windows of their mind to be opened to allow entry of light and winds from the outside world. The young students not only needed to do their own studies, they also needed to know the world around them and to take their own decisions about their own lives.

In order to give a boost to education, we constructed students' hostels to encourage the students, especially the girl students, to come out of the villages and to take admission in higher-secondary schools in the nearby towns. We had already set up an English-medium residential school near our steel plant project at Nagarnar. The purpose was to impart quality education to tribal children. But then, the task which we had taken upon ourselves was huge. We came to know of an excellent work being done in Bhubaneshwar, where one single man was running a residential school with 12,000 boys, mostly tribals, from a single campus. I felt that if we had to really bring about a change, we should do something similar to it in Bastar. I went to Bhubaneshwar and visited the school myself. Not only was education free in that school, the students were not required to pay anything for their board and lodging. Instead, the students were allowed to do some work for a couple of hours every day, and whatever items such as small paintings, bags or decorative pieces they could produce were marketed by the school and the proceeds went to the families of the concerned students. I requested the founder of the school to build and operate a school of similar type in Bastar, promising that all expenses on construction and management of the school would be borne by NMDC. The state government agreed to provide land and we could start the project work by laying the foundation stone for the project before I left the company.

I observed one interesting thing. Despite so much of violence prevailing in the area, our volunteers who went into the remote villages never faced any problem from anyone, including the armed insurgents. On one occasion, after visiting a school, the group observed that one student was by mistake not given the scholarship and was not attending the school either. It felt that just because the student was not given scholarship, he might have faced difficulty in attending the school, so they decided to go to the boy's house and talk to him. When they entered the boy's house, they found that their assumption was wrong; the boy had actually gone out of Bastar for a few days and just returned. They told the boy to fill up the required form and draw his scholarship after a week or so from the post office. Meanwhile, some people from the neighbouring houses came down to the boy's house to observe what was happening. They were stunned. They told our volunteers that they could never imagine that some people like them could come to the village and even to the house of a student only to ensure that the boy got his scholarship and that he faced no problem attending his classes. They persuaded the volunteers to have dinner in the village along with others, and when they were allowed to leave, it was already about eight in the evening. Some of the volunteers were internally anxious, but they soon realised that a relationship of trust had already been built between them and the local people and no harm would come to them. I remember having met with them a few days thereafter and heard of their experience, which touched not only their hearts but also that of mine.

Our work in Bastar, especially on education, created a positive impact not only on the local people who, despite their apparent backwardness and poverty, had an intense desire that their children acquired good education, it also influenced the state administration, which came forward in a big way in building coaching centres, schools and hostels, especially girl's hostels, to encourage the students of the region not only to pursue their studies but also to look forward to being professionals in different fields. I remember that the Chief Minister of the state at the time of inauguration of the residential school near the steel plant had clearly announced that the state would follow the example of NMDC and expand education in the state in order to enlighten, empower and enrich the local youth with knowledge, skill and opportunities so that they could not only pursue their dreams for better quality of life but also get integrated with the wider world. I had the opportunity to visit some of the hostels and training centres set up by the state government, and I must admit that those were immensely beneficial for the local especially the tribal people.

Our health care activities, which were initially covering 58 villages around mines in Bailadila, were extended to other locations, especially in the area where the steel plant was coming up. In Bailadila, the facility for hospitalisation was provided in the company's hospitals, one of which was a super-speciality hospital run by a renowned group. However, we also created facilities for the transfer of patients to even better hospitals in the state capital or in Hyderabad so that the best medical care could be provided. We created a team of medical professionals near our steel plant site to extend medical facilities to the local people in addition to providing treatment at the patients' door step by deploying mobile medical vans. We developed a tie-up with the state-run medical college for treatment of local people in case of necessity. We felt that medical care was needed for immediate relief for the bare survival of people; education was needed for enabling the people right from the young age to develop the right skill, right attitude and knowledge so that they develop eligibility; and jobs were necessary for ensuring a stable life with superior quality. We knew that once a person gets a job or acquires a way of livelihood, the need for initiatives toward health care and education would gradually go down, and that was the final aim of our work on community development. Our steel plant, meanwhile, was coming up in a big way to realise our dream of providing livelihood to hundreds and thousands of people directly and indirectly. We felt that what we were creating was not only an establishment for making steel but also an opportunity for the local people to look forward to having better lives for themselves and for their future generations.

NMDC was involved in CSR work for many years. My effort was to renew and to strengthen the activities further. The outcome was astounding. I believe that three factors contributed in this. First, we could develop the right strategy. Our idea of reaching out to people directly and spending money for individual benefit rather than development of local infrastructure had very good impact. Second, we were never hesitant in spending money on CSR. We always felt that the money spent on CSR was not only for a good cause, it was also our obligation to stand by the people to whom we owed quite a lot. As a company with strong fundamentals and comfortable profit, we could also afford to spend money on CSR without hesitation. Third, we and our people had no dearth of sincerity in carrying out the CSR work. To a large extent, we did our CSR work ourselves. We sincerely believed that working for the community and for the people was important and we should be directly involved in that work without involving other agencies as far as possible. This developed a direct touch between the company and the beneficiaries of the CSR work. It opened up opportunities for direct dialogue, and all these ultimately gave us the desired results. The government department which prepares guidelines for the public sector companies had issued a circular specifying the type of work which the public sector companies under CSR should carry out. After a few months, a new circular came up clearly stating that other companies should rather follow the example of NMDC.

I would like to end our CSR experience by mentioning a small incident. A reputed director was shooting a film in some villages near our mining establishments in Bailadila. I had an opportunity to meet him after my superannuation. By then, he had completed his shooting. He was describing his experience, especially the co-operation he got from the local people. I was curious to know how he could confidently take his team inside the forest and how he could secure the co-operation of the people in that area. His answer was simple. He had made all his crew members wear NMDC's uniforms and that gave him and his team all the protection and cooperation that he received. There could not be any better words to define the success of our CSR work.

8

LESSONS TO REMEMBER

Unlike Hindustan Copper, NMDC had been a strong company with strong fundamentals, a good work culture, good assets and a reasonable profit. While taking the company forward on the path of growth and diversification, I gathered some very useful experiences, which I have shared in the preceding chapters. The stories I have told are those of the successful turnaround of an immensely strong company, which was like a sleeping tiger that finally woke up to realise that it had all the powers and strengths to find a place in the world that it deserved. The strategies and efforts which were used to make this happen, left a number of lessons to be learnt.

First, even for an otherwise successful company with adequate resources and good work culture, growth initiatives need to come from the top, and their success depends to a large extent on the leader and his/her direct involvement. Leadership is important not only for turning around a sick company but also for forward movement of a strong organisation. A good team supplements good leadership but cannot supplant it.

Second, an organisation's excessive dependence on the leader for its turnaround, in whatever form it could be, is inherent in the organisational structure of a company. A company, especially a large company, builds an organisational structure on the basis of division of labour at the lower level and functional specialisation at the higher level of functioning. Even up to the level of director or board-level executive, the functionaries represent certain functions or areas of the operations and do not generally develop a holistic approach. The job of entrepreneurship rests at the level of the CEO or the leader of the organisation and, therefore, visions and long-term growth initiatives flow from the CEO to the lower level and rarely in the reverse direction.

Third, resistance to change is inevitable. Any initiative which is associated with change, including minor changes, needs to be taken care of well before the initiatives are taken as otherwise, with every forward movement in any developmental initiative, the degree of resistance will grow in multiplication. In NMDC, we were fortunate to get the support of a strong federation of unions and we opened up communication with it well in advance, before any change was initiated.

Fourth, there is no substitute to direct communication with people. When we decided to set up the steel plant and held a public meeting, we decided to do two things. I as CMD took the task of communicating directly with the local people upon myself. I knew that any commitment made by anyone other than the CMD would not be trusted by the people. Also, it was ensured that there should be two-way communication in which a good amount of time was allotted for the people to respond, to speak and to come out with their views. This approach was proved to be extremely beneficial in the times to come.

Fifth, rules and regulations are useful in a company, but one must remember that these are nothing but codified experiences. A vibrant management always adjusts its rules and procedures with the requirements of the organisation and the realities of the emerging situations. A good management knows that rules are not there to restrict but to empower an organisation and facilitate its functioning. Common sense and confidence are two essential ingredients for developing new approaches, strategies and rules. Once an organisation proceeds to develop a new approach on the strength of its own logic and conviction in deviation from the traded path, it becomes a path-breaker and acquires a new relevance in its existence. This is exactly what happened in the case of NMDC.

Sixth, whatever is good for the nation and good for the society is also good for the company. Whenever the nation needs something or the society aspires to have something, a commercial organisation has to see opportunities in them. These are actually essential inputs for a company which intends to grow and looks for possible areas for expansion. An organisation's profit comes from value additions, and the value it derives for its products and services depends on the need that the society has for them.

Therefore, national interest and organisation's interest are not in conflict. Further, while deciding to invest in response to the aspirations of the society or the community, an organisation should not be bound by the concept of core competence. Core competence is more relevant for a person than for an organisation. A person has to struggle to develop a skill beyond the core competence, but an organisation can step out of core competence just by hiring people with the relevant skill and competence. When we understood that we had to invest for building a steel plant, many people told us that it would be beyond the core competence of NMDC. All of them were proved wrong when NMDC became the fastest in acquiring land, placing orders and carrying out the construction work.

Seventh, no expansion should take place at the cost of core competence. When a company expands into a new area of business, under no circumstance should the existing business in which the company has proven capabilities should suffer. When we decided to set up the steel plant, we were very careful to ensure that our mining business should continue to be strong, our customer base should not be disturbed and our investment in mining and in protection of mineral resources should continue to be carried out in the same way in which it was done in the past. We insisted that we should get a separate mine to feed our steel plant. If any iron ore was to be supplied to the steel plant, it should be done at market price. We, rather, took the steel plant as an opportunity to dispose of our mining wastes and minimise the supply of marketable material to the steel plant. While investing in the steel plant, we also wanted to ensure that our plans to invest in the beneficiation of low-grade ore and tailings, in setting up facilities for pelletisation and in carrying out intensive exploration within the mining area were executed as quickly and as seriously as our construction work was being carried out in our steel plant project.

The eighth lesson has been that the most important investment of an organisation is investment in people, not only those who are working for the organisation but also those of the community whose lives and aspirations to a large extent depend on an organisation, which influences them directly or indirectly.

Expenditure on CSR as a statutory obligation defeats the very spirit of CSR work. Companies need to carry out CSR work on their own initiative to ensure that they exist not only for themselves and their employees but also for the community around them. Further, I always found that direct work of company on CSR gave much better result, not only for the community but also for the company itself.

Ninth, when an organisation grows, the value of participation of employees at every level is felt more intensively.

In NMDC, the quality circles proved to be very useful participative forums. What they needed was more support and more attention. Active participation of workers and officers in analysing the problems and opportunities at the organisational or working level keeps the fundamentals of the company strong. If the hand which drives the machine is driven by an active mind, the company gains and the worker derives such motivation which no other system can provide. NMDC had one more advantage. Most of the employees lived in the company's townships near the mines which were located in remote areas. Their isolation from the outside world brought them more close to each other and encouraged them to be more involved in the affairs of the company. As a result, communications made to one section of employees percolated down to others, and together everybody not only deliberated on the subject but also came back with their feedback. The company gave the employees an identity, which they were proud of, and this made the groundwork for more meaningful participation of employees at all levels. It was most common in Bailadila that in the months of January and February, when the area experienced severe winter with biting cold, the housewives would push their husbands out of bed lest they miss their vehicles for attending work in the first shift because in these months the employees had to meet very stiff targets.

Lastly, when an organisation grows, it is important not only that its activities expand or that its financial position improves but also that the organisation acquires new importance, new acceptance and new significance in the national and international arenas. Any growth is usually reflected in increase in business. Improvement in the top and bottom line comes as a consequence. An organisation's urge to grow is, however, not satisfied with these achievements; it aspires to create a socially accepted value, which goes beyond the financial achievements. These achievements relate to the recognition it gets from national and international levels, its capacity to influence policy-making at the national level and the significance it enjoys at industry level or in the society. A company grows when its opinions, judgements and achievements are valued by the wider society, which treats it as a pathfinder. NMDC's achievement had been that, if not fully but at least substantially, it could elevate itself to that stage.

When I left Hindustan Copper in 2005, I carried a satisfaction that I could contribute to bringing a company back from the brink of disaster; with my humble efforts and dedication and luck, the company could not only survive but could become a profitable company after many years of sickness. I thanked destiny for giving me an opportunity to lead the entire struggle from the front. NMDC did not have any dearth of resources. I had only to decide how the strengths of the company could be used to realise its full potential. It was a more difficult challenge and, of course, more complex. When, during my tenure, on three consecutive years NMDC was rated to be the best performing PSU, best performing Navratana company and best performer on CSR, I felt that my goal was largely achieved. When I found that NMDC's examples were being followed at the state and national levels, its policies were adopted as national policies and its opinions were valued at the highest level, I realised that I could reach very close to achieving the mission which I had set for myself in NMDC. I left my office on the last day of 2011 and still carry the satisfaction with which I superannuated. This possibly has been the inspiration for this book.

"In this book Pradip Chanda has distilled years of personal experience into powerful lessons for leading such a turnaround. Every senior business leader would benefit from reading this book."

Sunil Gupta
Edward W. Carter Professor of Business, Harvard Business School

Powerful Lessons for Scripting Turnarounds

For special offers on this and other books from SAGE, write to marketing@sagepub.in

Explore our range at
www.sagepub.in

₹395

Paperback
978-93-860-6242-0